Asymmetries in Time

Asymmetries in Time

Problems in the Philosophy of Science

Paul Horwich

A Bradford Book
The MIT Press
Cambridge, Massachusetts
London, England

This book was set in Bembo by Asco Trade Typesetting Ltd., Hong Kong, and printed and bound by Halliday Lithograph in the United States of America.

Library of Congress Cataloging-in-Publication Data

Horwich, Paul.
  Asymmetries in time.

  "A Bradford book."
  Bibliography: p.
  Includes index.
  1. Science—Philosophy.   2. Space and time.
  3. Causality (Physics)   4. Entropy.   I. Title.
  Q175.H796   1987   501   86-28632
  ISBN 0-262-08164-4
  ISBN 0-262-58088-8 (pbk.)

Q
175
.H796
1987

To my brothers
Alan and Roger

Contents

5
Knowledge

6
Backward Causation

7
Time Travel

8
Causation

9
Explanation

Preface

Time is generally thought to be one of the more mysterious ingredients of the universe. Perhaps some of the reason for this is that *understanding* is often a matter of finding analogies. But time is unique; there's nothing else remotely like it. Another, more tractable, source of perplexity is that the notion of time interacts in subtle and convoluted ways with a great variety of the deepest ideas in our conceptual scheme. It is this jigsaw puzzle that I am struggling with in the present essay. I try to make precise and explicit the interrelationships between time and a fair number of philosophically important ideas. To the extent that this is successful, time may seem to be not so mysterious after all.

This book is about phenomena, such as knowledge, causal influence, and entropy, that are biased with respect to time. Its primary focus is on questions like: Why is the direction of time from past to future? What is causation, and can it work backward? Could an act be rational if performed for the sake of a desirable past state? How is it that we know so much history, yet so little about the future? Can one properly explain the characteristics of a system in terms of its goals, rather than its origins? Why is there a profusion of decay processes but little spontaneous generation of order? And, finally, are there deep relationships among these temporally asymmetric phenomena? My principal thesis is that the asymmetries are indeed closely interconnected, and to sustain this claim, I offer a detailed picture of the explanatory links between them.

I shall be dealing with a constellation of notoriously problematic concepts—cause, knowledge, explanation, entropy, rational decision, law of nature, and counterfactual implication—and my secondary aim is to help clarify these ideas. Certainly there can be no presumption to have given complete, definitive accounts in the single chapters allotted to those topics. There can, however, be the hope that our concentration on time asymmetry, and the global treatment of all the ideas in relation to one another, will provide a simplifying focus without loss of breadth and thereby enable us to make some progress toward a deeper understanding of each concept.

Readers who want an overview of the issues should begin at the beginning; but otherwise, because the chapters are each fairly autonomous, they can easily be taken out of order. The first half of the book concerns time more explicitly than the rest of it does. Thus, following the introduction (chapter 1), there are chapters devoted to the metaphysics of *now* (chapter 2), the meaning of the thesis 'time is asymmetrical' (chapter 3), entropy growth (chapter 4), how it is that we know so much about the past (chapter 5), and the possibilities of backward causation (chapter 6) and time travel (chapter 7). In the second half I turn to more familiar problems in the mainstream of philosophy, but considered from a temporal perspective. In particular, there are analyses of causation (chapter 8), explanation (chapter 9), counterfactuals (chapter 10), and rational choice (chapter 11) and, finally (chapter 12), a sketch of results and residual difficulties.

Although this work is in the philosophy of science, very little previous knowledge of science is presupposed, and there is no danger of anyone's being blinded by technical material. What little physics there is, is almost wholly confined to chapters 3 and 4, on anisotropy and entropy, and occasional sections elsewhere. These parts are self-contained, informal, and accessible to a nonspecialist. Moreover the rest of the book does not heavily depend on them.

The philosophical tradition to which this essay tries to contribute, began in earnest with Hans Reichenbach's classic work, *The Direction of Time*, published posthumously in 1956. This was the first attempt to give a complete account of temporal asymmetry. It was followed in 1963 by Adolf Grünbaum's *Philosophical Problems of Space and Time*, which endorses Reichenbach's views on most major points but provides a great deal of vital clarification and argument. Subsequently, a series of papers by John Earman (especially "An Attempt to Add a Little Direction to 'The Problem of the Direction of Time'" 1974) has brought to light some powerful objections to the Reichenbach/ Grünbaum approach, and has helped to make clear what an adequate account must do. And in the last fifteen years, David Lewis has written half a dozen articles on counterfactuals, causation, and decision (collected in his *Philosophical Papers*, Vol. 2, 1987) which add up to a formidable theory of temporally asymmetric phenomena.

I hope that no one is offended by the unceremonious attitude I am going to take toward these pioneers of the subject. Despite the many disagreements I well know how much this work owes to their insight and example. It has also benefited from contributions by many others, and I am delighted to thank these people. Susan Brison talked with me about large parts of the project, and gave friendship and encouragement. Ned Block and Josh Cohen commented on all the drafts of every chapter, constantly pushing me to try harder. I hate to think what this book would have been like without their generous and painstaking criticism. Judith Thomson applied her high standards of clarity and rigor, and helped me eliminate many obscurities. Jeremy Butterfield, Hartry Field, Sidney Shoemaker and Bob Stalnaker provided valuable detailed reactions to an early draft of the entire manuscript. Izchak Miller spent hours explaining to me Husserl's theory of our awareness of the passage of time. David Malament corrected several errors in my discussion of Gödelian time travel. Tom Kuhn and Larry Sklar did the same for the chapter on entropy. Mike Williams and John Carriero got me to see the force of Hume's ideas on causation. And I have had useful and enjoyable discussions with many other people, including Sylvain Bromberger, Jay Cantor, Ellen Eisen, Marcus Giaquinto, Bernard Katz, Fred Katz, Jerry Katz, and Abner Shimony.

Two parts of the book have been published before, and I am grateful for permission to reprint them. Chapter 7 is an improved version of an article, "On Some Alleged Paradoxes of Time Travel," which appeared in the *Journal of Philosophy* in 1975; and chapter 11, except for the final section, was published in *Philosophy of Science* in 1985, under the title "Decision Theory in Light of Newcomb's Problem." Finally, I would like to express my appreciation to the National Endowment for the Humanities and to the National Science Foundation.

Asymmetries in Time

1
Asymmetries

1. Aims

One of the most puzzling things about time is its flavor of asymmetry. We speak of time's "arrow" and "flow", trying to capture this feature—intending to suggest a profound difference between past and future. However, it isn't easy to do without these metaphors—to say literally what is meant by supposing that time has a direction in order to settle the question of whether or not it really does have one. Nor is it clear what we should make of pervasive temporal asymmetries like our capacity to control the future, the prevalence of decay, and the relative ease with which we can obtain knowledge of the past. Such asymmetries in time call for explanation, but do they indicate that time *itself* is asymmetric?

The atmosphere of mystery surrounding 'the direction of time' is crystallized in many specific philosophical and scientific problems. Let me mention three examples, and then give a more systematic account of the issues before us.

An especially intriguing problem concerns time travel. Is it possible to 'go back in time'? If so, how could it be done? And if the necessary technology will ever become available, why haven't we yet encountered visitors from the future? These questions are particularly tantalizing, as the topic of time travel is no longer confined to the realm of pure fantasy, having gained a measure of scientific respectability in the work of Kurt Gödel. It had been known since the acceptance of the Special Theory of Relativity that a form of time travel into the *future* could be accomplished by exploiting the fact that moving clocks run slowly. Thus someone could go on a very fast rocket trip and age, biologically, only two years, yet find on his return that ten years of Earth time had gone by. This sort of thing may or may not be properly called "time travel". In any case its possibility is uncontroversial. Much more problematic, however, is the idea of time travel into the *past*. Gödel's striking contribution was the discovery of certain spacetime structures, consistent with the General Theory of

Relativity, that would allow journeys into history: ". . . by making a round trip on a rocket ship it is possible in these worlds to travel into any region of the past, present and future and back again, exactly as it is possible in other worlds to travel to distant parts of space" (1959, p. 560). But is this really conceivable? For how many of me would there be if I went back to 1950? Could I shake hands with myself? Could I change history—do something that in fact was not done? Worse still, what would stand in the way of 'autofanticide'—killing myself as an infant—an action whose failure would seem to be a necessary condition for success? We are left, therefore, with the following general question: Must we dismiss Gödel's results as mere mathematical curiosities, or can we dissolve the apparent paradoxes and preserve the possibility of time travel?

A more mundane problem involving time asymmetry concerns human motivation. What is the logic of rational action? What factors make one of a person's choices more reasonable than his other alternatives? This question interacts with the concept of time direction in that our processes of deliberation are *future* oriented. We try to figure out what we can do to maximize the chances of future benefits, and we don't much worry about the past, except perhaps for bouts of regret and self-congratulation. But why is this? Why is deliberation biased with respect to time? The explanation might seem obvious: namely, that causal influence works toward the future. We can never affect the past, so there's no point in planning for it. But there is another more subtle answer that cannot be dismissed. In the course of deciding what to do, we entertain various options. We find that our beliefs about what *will* transpire *vary* along with these different suppositions, but our conditional beliefs about what *has* happened remain constant. So, as far as the past is concerned, how can one action seem preferable to another? Thus our 'conditional belief asymmetry' suggests an alternative explanation of why past-oriented desires do not motivate us—one that does not invoke the direction of causation.

This pair of apparently conflicting answers to the question of why we act only for the sake of the future reflects two general conceptions of what makes something worth doing. On the one hand, there is the familiar, prevalent view—that the choiceworthiness of an act stems from what it might *cause*. On the other hand, and more liberally, we might take into account any event that the act would be *evidence* for. That is to say, even those events that are *not* caused by the act, but merely correlated with it, may be said to contribute towards its choiceworthiness.

Now one might wonder if there is any real dispute—or even any real difference—between these conceptions. But the answer is yes, and Newcomb's notorious decision context clearly exposes their divergence. Imagine that a superior Being is able to foretell human choices, reliably and accurately, by means of a fancy psychological theory. Now suppose you are informed that yesterday the Being calculated whether or not you will now scratch your head, and that he set aside for you a valuable gift if and only if he predicted that you would scratch. Do you have good reason to do so? The causal conception of motivation says no. For it is not the act of scratching—but rather its precursors, recognized as such by the Being—that would cause the benefits. The evidential theory says yes. For given one's knowledge of the Being's abilities and intentions, scratching does raise the probability of benefits, even though it cannot bring them about. So the theories disagree. And pretheoretical intuitions don't settle the matter; they differ wildly from one person to another. Thus we are faced with three connected questions: Which is the right account of rational choice? Which action should be performed in Newcomb's dilemma? And which explanation of the decision time asymmetry is correct—why do we act for the sake of the future and not for the sake of the past?

These sample problems are *philosophical* insofar as their solutions involve clarification, examining relationships between concepts, and the unraveling of confusion. But not all our puzzlement about 'the direction of time' may be treated philosophically. Some of it reflects the need for a better *scientific* understanding of temporally asymmetric phenomena. We want to know why entropy tends to go up—why, in other words, highly ordered states decay but do not spontaneously evolve. This sort of asymmetry is commonly said to be, or to result from, the "arrow of time". But perhaps it has nothing at all to do with time. Perhaps time itself is perfectly symmetric, and increasing entropy is caused by cosmological conditions that do not bear on the nature of time itself. Moreover we are interested in what *follows* from the profusion of decay processes, and not just in the conditions that bring about this phenomenon. For example, it would be surprising if our impressive knowledge of the past (compared to our vast ignorance of the future) were unrelated to those physical time asymmetries. In addition there is the future-directedness of causal influence to account for. Many philosophers have thought that the scarcity—perhaps impossibility—of backward causation is another deep effect of entropy growth.

The purpose of the following study is to investigate the entire cluster of questions—including those just mentioned—often lumped together under the heading "the problem of the direction of time."

The project will involve (1) a clear specification of what it is for time to be asymmetric (directed, anisotropic), (2) a precise characterization of several pervasive temporally asymmetric phenomena, such as the fact that we are able to influence the future but not the past, and the fact that we know a great deal more about the past than the future, (3) an attempt to explain these asymmetries in a unified way and to examine their bearing upon the directedness of time itself.

In what immediately follows I shall describe ten apparent asymmetries that provide the main stimuli for our inquiry. Then I shall indicate the sort of understanding of these phenomena that will be our goal in subsequent chapters.

2. Ten temporally asymmetric phenomena

Now

We have a sense that time flows. We recognize a one-dimensional continuum of instants at which events are temporally located. But in addition there seems to be a kind of gliding index—*now*—that gradually moves along this array in the direction from past to future. "Time is the moving image of eternity," said Plato. "It is as if we were floating on a river, carried by the current past the manifold of events which is spread out timelessly on the bank." Or as Ovid put it, from a different point of view, "Time glides by with constant movement, not unlike a stream, for neither can a stream stay its course, nor can the fleeting hour."

This idea is captured less metaphorically in McTaggart's (1908) widely held theory about what would be required for the existence of time. According to McTaggart, the world contains a sequence of events ordered by such relations as *later than* and *simultaneous with*. But this would not suffice, he thought, for time to be real. In addition, to provide for genuine change in the universe, there must be a series of temporal specifications—distant future, near future, now, past, and so forth—with which events may be located, and which slides along the sequence of events in such a way that *now* applies to continually later and later events. McTaggart believed that, although essential to the existence of time, this 'motion of *now*' would be self-contradictory (since every event would have to possess the incompatible attributes of being past, present, and future). He concluded therefore that time is unreal.

This conclusion is literally incredible. Nevertheless, there is much to be learned from McTaggart's line of thought. We might well be favorably impressed with just the second component of his argument. We might agree (though this will take some showing) that the 'moving *now*' conception of time is indeed incoherent. But we might suppose this to mean, not that time is unreal, but that time need not (indeed *cannot*) be associated with a 'moving *now*'. If this is correct—if the 'moving *now*' is really just an illusion—then a further problem begins to loom. Why does this illusion have such a hold over us? Why is there such a strong sense that time does pass in a particular direction if it really doesn't?

Truth

Following Aristotle, it is often maintained that contingent statements about the future have no truth value, unlike statements concerning the past and present which are determinately either true or false. A prediction that war will break out next year will attain a truth value only then, when the event occurs or fails to occur. But right now there is no fact of the matter; for if there were, the presence or absence of the war would now be fixed, and nothing could be done to influence it. This contention is intended to imply a 'tree model' of reality whereby the past is petrified, uniquely fixed, over and done with, but the future contains a branching manifold of undetermined possibilities. And there is a tendency to invoke the 'moving *now*' in order to explain how a definite path up the tree is selected. It was Aristotle's belief that such an ontological distinction between the future and the past would have to exist if determinism and fatalism are to be avoided. We must assess this view and understand more precisely its relationship to the 'moving *now*' conception of time.

Laws

Turning now from metaphysics to science, suppose there is a process whose temporal mirror image is impossible—that is, a sequence of states of affairs, *ABCD*, where the reverse sequences, *DCBA*, would be ruled out by laws of nature. In such a case *ABCD* is said to be nomologically irreversible, and the operative laws are said to be time-asymmetric. Most of the theories that have ever been seriously entertained are not of this sort. They are like Newtonian mechanics in being time-symmetric and permitting the temporal reverse of all the processes in their domain. For example, imagine the motion of billiard balls colliding with one another on a frictionless table. If a movie of such a process were shown in reverse, no violation of law would be apparent. On the other hand, consider the second law (so-called) of thermodynamics, which states that the entropy of an isolated system will never decrease. This is evidently not time-symmetric, as there are processes of entropy increase (e.g., milk spreading through a cup of coffee) whose temporal inverses are ruled out by it. Although thermodynamics has been superseded, the prospect of nomological irreversibility has not been banished from science. There is evidence that certain forms of fundamental particle decay (of the neutral K meson) are nomologically irreversible.

Our study of the asymmetry of time will involve an investigation of several questions provoked by the concept of irreversibility. Why, for example, is it so readily taken for granted that irreversible phenomena guarantee the asymmetry, or anisotropy, of time itself? This thesis will have to be justified in light of a clear statement of what it is for time to be asymmetric. Also, what is meant by the "temporal mirror image" of a process? This isn't at all obvious. One cannot simply say, as we have just blithely assumed, that the mirror image of *ABCD* is *DCBA*—the very same constituents in reverse order. For imagine a man eating dinner: soup, meat, desert, and then coffee. Must we hold that the reverse process is simply a meal in which coffee comes first? Then there is the matter of 'direction'. Would the presence of time *asymmetry* have any bearing on the so-called '*direction* of time'? Why do we single out one of time's two directions for the title "*the* direction of time"? And is it conceivable that time should change direction?

De facto irreversibility
There are many processes whose temporal inverses are possible, although in actuality they never, or hardly ever, occur. Schematically the sequence of states *ABCD* is common, whereas instances of *DCBA* are extremely rare. For example, if a gas is concentrated in some small part of a container, it will expand to fill up the whole space available to it. But a gas that initially occupies the whole of its container never spontaneously shrinks into one corner. Similarly a source of light will emit a spherical beam that radiates outward; however, it never happens that a concave spherical beam converges toward a single point. In such cases we are dealing with *de facto* one-way processes whose inverses don't happen, though they are not precluded by the laws of nature.

Evidently some of the very problems arise here that I have just mentioned in connection with nomological irreversibility: problems that concern the meaning of "temporal inverse" and the conditions for the asymmetry of time itself. We shall examine arguments, on the one hand, that mere de facto irreversibility suffices to confer anisotropy upon time (Grünbaum 1963) and, on the other hand, that the association of irreversibility with the directionality of time is a mistaken dogma (Earman 1974). However, the central problem presented by these de facto one-way processes is to *explain* the temporal asymmetry that they display. It will be seen that laws of nature are not enough to account for it, especially given Boltzmann's reduction of thermodynamics to statistical mechanics. This means that the high frequency of entropy-increasing decay processes must be attributable to certain de facto conditions of the universe. Our aim will be to identify these conditions and to find their cosmological basis.

Knowledge

We know more about the past than we know about the future. For example, it is much easier to describe yesterday's weather than to forecast tomorrow's, and we know when the last five earthquakes in California took place, but not when the next five will be. Admittedly our capacity to predict in certain areas is quite impressive, and there are huge gaps in our knowledge of history; so it is hard to give a precise characterization of the asymmetry without exaggerating it. Nevertheless, it seems undeniable that there exist in the present many traces of earlier circumstances but relatively few reliable and recognizable indicators of what is to come. Thus we must acknowledge a dramatic difference in epistemological accessibility between the past and the future regions of time.

In trying to explain this time asymmetry, we shall look at various possible causes of it: (1) our capacity to act freely, and thereby to affect the future unpredictably; (2) our inability to go back in time and verify historical claims, which allows us to think mistakenly that our knowledge of the past is superior; (3) the direction of causation, which enables causal traces of the past but none of the future; (4) again, the direction of causation, because the meaning of the word "know" prohibits knowledge of any occurrences that do not cause our awareness of them; (5) the fact that, although the future may be physically determined by present events, the past is physically *over*determined, so current conditions provide several independent determinants of what has happened; and (6) the second law of thermodynamics, which allows us to infer, given the observation of a highly ordered system, that it has previously interacted with its environment. We shall find that none of these explanations is good enough. My analysis will proceed by examining the general nature of actual recording systems, such as books, photographs, and memory, that give us our special knowledge of the past, and then asking why it is that the temporal mirror images of such systems do not occur.

Causation

Effects seem never to precede their causes. We can influence the future but not the past. In other words, backward causation is impossible, de facto nonexistent, or, at the very least, extremely rare in this part of the world. The point of my weak characterization of the phenomenon is to avoid controversial presuppositions. For even if backward causation *sometimes* occurs, it remains to be explained why the *predominant* direction of causation is toward the future.

It is tempting to say: "A cause is, by *definition*, earlier than its effects." But this account must face up to several objections. Does it not trivialize the absurdity of acting for the sake of past events? For can we really suppose that the extreme irrationality of such a retroactive policy is merely a matter of stipulation? And what about cases of simultaneous causation? There appear actually to be such cases, yet they would straightforwardly falsify the alleged definition. Moreover is it right to consign backward causation to the same realm of inconceivability as the married bachelor or the weekend in which Sunday comes before Saturday? After all, we are able to *imagine* circumstances in which it might be tempting to say that an effect has occurred before its cause. To accommodate these difficulties, we shall have to take seriously certain alternative analyses of causation, accounts whereby its directional character does not stem purely from definition but depends on various contingent features of the world.

On the question of whether an effect can *ever* precede its cause, the issue turns largely on the merits of a well-known line of reasoning: namely that any backward causation hypothesis would automatically be refuted simply be waiting for an occasion on which the alleged effect is not present and then producing the alleged cause. I will try to show that this objection—sometimes called "the bilking argument" —is not as powerful as it is often taken to be. And I will apply this conclusion to three physical theories that postulate backward causation. Specifically I shall assess its effect on the plausibility of superluminal signals (tachyons), Feynman's identification of positrons with electrons moving backward in time, and Gödelian spacetime. There is plainly a close similarity between the bilking argument and some of the paradoxes of time travel that I mentioned at the start.

Explanation
We seek explanations of phenomena in terms of antecedent, rather than subsequent, circumstances. Indeed, explanations are sometimes roughly defined as accounts of an event that show how it could have been predicted—how, given the prevailing conditions, it was only to be expected. On the other hand, so-called teleological and functional explanations, which purport to explain a system's present state in

terms of the attainment of some future goal, are of dubious scientific respectability. Why did the chicken cross the road? Presumably because of its earlier condition of wanting to reach the other side and not because of its later state of being there. Why is space three-dimensional? Some physicists (adverting to what is known as the Anthropic Principle) say the reason is that, otherwise, stable planetary orbits would be impossible, and so life—and our awareness of the three-dimensionality of space—could not have evolved. But this sort of account, unsupplemented and taken at face value, can easily strike one as fundamentally misconceived, blatantly putting the cart before the horse.

It is plain, I think, that our ideas about the proper direction of explanation are intimately bound up with the directionality of cause and effect. What is not so clear, however, is precisely how these phenomena are related. A natural answer is that *explaining* is a matter of *specifying causes* and consequently that the explanation asymmetry is a product of causal directionality. But there is an interesting alternative to this approach. Reichenbach (1956) and Dummett (1964) have expressed the view that the arrow of explanation stems from the prevalence of de facto irreversible processes and that the direction of causation then derives from our concept of *cause* as *explainer*.

Counterfactual dependence

If the present were different from the way it is, then the future would be different. Thus true counterfactual conditional statements of the form 'If A had not occurred, then C would not have' seem to be about what would have happened subsequently if some actual event had not taken place, and not about what would have happened before.

For example, suppose the actual facts in some situation are (1) that Jones was on the roof of a high building, (2) that there was no safety net underneath him, (3) that he did not jump, and (4) that he was not hurt. A hypothetical negation of 3—the supposition that Jones jumped—would be taken to imply the negation of a subsequent fact, 4. In other words, we accept

If Jones had jumped, he would have been hurt

On the other hand, that same supposition is not easily taken to imply an alteration in preceding conditions. We do not normally conclude from the fact that Jones was on a high building and did not get hurt

If Jones had jumped, then there would have been a safety net underneath him

However, having noted the apparent time asymmetry of counterfactual implication, we must acknowledge that the issue is far from clear. One can quite well imagine someone saying

Jones would have jumped, only if there had been a safety net

And does this not mean

If Jones had jumped, there would have been a safety net

which implies

If Jones had jumped, he would not have been hurt

contradicting the initial claim?

In order to sort out this tangle of conflicting intuitions—and to determine whether such conditionals are really time-asymmetric —we must clarify the meaning of the counterfactual "If. . . then. . . " and specify its inferential relationships to other concepts. In particular, *causation* is a closely affiliated notion, and it will be important to understand precisely how it interacts with counterfactual dependence. The leading current approach to these matters is due to David Lewis (1973a, 1973b, 1979b), who argues with great flair that counterfactual dependence *is* time-asymmetric and that the direction of causation is an immediate consequence of this phenomenon. However, as we shall see, Lewis's approach is afflicted with a multitude of serious difficulties. It will be worth exploring the possibility that his theory inverts the actual explanatory relationship, and that counterfactual dependence grows out of the causal/explanatory order, rather than vice versa.

Decision

We act for the sake of the future, not the past. More precisely, we would think it gravely irrational for someone to do something in order to ensure, or make probable, the occurrence of some desirable past event—or to preclude an undesirable one. It's no use crying over spilt milk, we say. In contrast, the future appears to be substantially controllable, and it is often reasonable, when some event of significance to us is in question, to be guided by whether any of our alternative actions would seem to make probable its occurrence.

One very natural explanation of this asymmetry is that we think it rational to act only for the sake of things we might *cause* to occur, and we are well aware that events, including our own actions, can exhibit causal influence only over the future. However, it is possible to resist this so-called *causal* decision theory. Instead, one might argue that the asymmetry results from the fact that our actions are probabilistically independent of prior circumstances. As I said at the outset, the conflict between these points of view comes to a head in the context of Newcomb's problem, and so this will be the focus of our discussion.

Value

We care a great deal more about what *will* happen to us than about what *has* happened. As a consequence we would much rather have pleasant prospects ahead and bad times behind us than the other way around. In particular, we dread death—future time at which we will not be alive—yet are quite unperturbed by the corresponding fact about birth—the reaches of past time when we were not alive.

This bias toward the future—the special importance we attach to what is still in store for us—has been noted by Derek Parfit (1984), who considers the question of what could account for it. I shall be concerned not only with that issue but also with the problem of how this bias is related to other temporally asymmetric phenomena. Is there, for example, any truth in George Schliesinger's (1980) thesis that the existence of this bias provides strong evidence in favor of the 'moving *now*' conception of time?

3. Explanatory maps

The general aim in the following chapters is to study each of these ten alleged asymmetries in order to discover: (1) which of them are genuine, (2) why they occur (3) what are the explanatory relationships, if any, between them, (4) whether there is some fundamental asymmetry (or two, perhaps) from which the others follow, and (5) what relation the asymmetries bear to the thesis that time itself is anisotropic (has intrinsic directional character). Are any of them constitutive of that thesis? Do they provide evidence for it? Do we need to invoke anisotropy in order to account for them?

To a large extent these objectives may be achieved by organizing the phenomena in a flowchart representing the explanatory relationships between them. As far as I know, the problem of the direction of time has not been approached before in this way and on such a broad front, and there is no explicit proposal of the sort of explanatory map I have in mind. Nevertheless, it is possible to extract from the writings of those who have struggled with these issues some partial theories. As an illustration I shall briefly describe and comment on the partial map (figure 1) implicit in Reichenbach's (1956) work.

According to Reichenbach, de facto irreversible processes consist in the creation and gradual decay of order. The observation of a highly ordered (low entropy) state tells us (given the second law of ther-

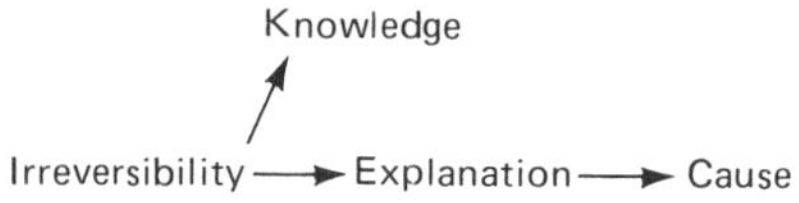

Knowledge

Irreversibility ——► Explanation ——► Cause

Figure 1

modynamics) that since the chances are negligible of spontaneous evolution into such a state, the system must have interacted in some characteristic way with its environment. Thus we obtain information about the past. In addition, by reference to the maximally ordered state, we may confer high probability upon the subsequent states of partial order—states that would otherwise seem amazingly coincidental and improbable—and thereby explain them. Thus the direction of explanation is tied to the direction in which order is dispersed. Moreover, assuming that our notion of *cause* is *that which explains*, we also derive the direction of causation indirectly from the orientation of irreversible processes.

This line of thought is extremely ingenious, yet controversial in every detail. For example, Mackie (1974) maintains that we should get the existence of irreversible processes from the direction of causation, rather than the other way around; Earman (1974) suggests that the knowledge asymmetry stems from the causal asymmetry, and that neither has much to do with entropy; von Wright (1971) says that the temporal orientation of causation comes out of our ability to manipulate the future; Salmon (1984) argues that explanation should be defined as a specification of causes.

Thus there is strikingly little agreement about the sources of temporally asymmetric phenomena and about the interdependencies among them. This is in some part because philosophers have tended to approach these questions in an overly piecemeal way. Consequently their conclusions are often undermined by a failure to appreciate and accommodate the needs of a comprehensive account. The following investigation will attempt to avoid this shortcoming. It will offer explanations of the asymmetries and criticism of alternative proposals. And its theses will gain credibility from their interaction within an unusually broad conceptual network.

The overall theory towards which I shall be working involves the idea that time *itself* has no intrinsic directionality or asymmetry, and it explains the temporally asymmetric phenomena accordingly, as shown in figure 2. To begin with, the de facto irreversible processes are given a cosmological explanation (in terms of the randomness of microscopic conditions following the big bang) and are employed to account for the fact that we know so much more about the past than

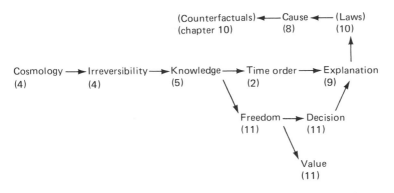

Figure 2

about the future. This knowledge asymmetry, by virtue of the difference between memory and expectation, yields our conception of succession. Time order is built into the notion of explanation in such a way as to imply that earlier facts may be taken to explain later ones. And the direction of physical explanation yields the direction of causation, since explanation is a description of causes. Counterfactual dependence is then analyzed in terms of causal explanation and, contrary to first appearances, it turns out not to be time-asymmetric after all.

Moreover an element in our sense of free choice is that the knowledge of what we are going to do evolves through a process of deliberation, intention, and action. Given the knowledge asymmetry, such a process must have a particular temporal orientation: namely, deliberation, followed by intention, followed by action. This ordering implies that our beliefs about what will occur in the future are sensitive to variations in what we suppose we will do. On the other hand, since the relationship between past events and our prospective action is mediated by the beliefs and desires that we recognize during deliberation, there is nothing we can normally infer about the past from some supposed action that we cannot already infer during deliberation independently of any such supposition. This is what gives rise to the decision asymmetry—our tendency to act for the sake of the future—which in turn plays a role in fixing the direction of explanation. The value asymmetry—the special importance we attach to future experiences—is also explained by the temporal orientation of our sense of freedom. Given that the typical decision process involves desire, deliberation, decision, action, and fulfillment (in that temporal order), then a desire for future satisfaction will be an aid to survival. For this attitude increases the chances that future selfish desires will be fulfilled; whereas there is no mechanism by which a present wish that past desires have been satisfied would be associated with any increased fulfillment of those past desires.

Needless to say, any sketch such as this is likely to be dangerously oversimplified. In particular, one should guard against supposing that the arrows in figure 2 always indicate the same type of explanatory connection. However, I hope, at least, that it is helpful to see at a glance the rough shape of the theory that will gradually emerge. A more detailed summary is given at the end of the book.

It is natural to think that time is obviously asymmetric, with an indefinable, yet undeniable, directional character, and to think that this is responsible in one way or another for the various asymmetries that we have just been discussing. But as I have indicated, my view is very different. The first step in presenting it will be to attack the idea that time has a direction (chapter 2), or indeed that it is asymmetric at all (chapter 3). Then, after discussing the behavior of entropy, I shall turn to an affiliated phenomenon, called "the fork asymmetry", which is closely related to the fact that regularly associated events must have a common cause but need have no joint effect. This phenomenon, I will argue, does provide a source for many of the asymmetries we are dealing with. The nature and limits of this dependence, and the role of other factors, are elaborated in subsequent chapters.

2
Direction

1. The 'moving now' conception of time

The quintessential property of time, it may seem, is the difference between the past and the future. And here I don't just mean that the past and the future are separate regions, or that the past and future directions along the continuum of instants are opposite to one another, but rather that these two directions are somehow fundamentally unalike. This idea is fostered by the desire to explain pervasive temporally asymmetric phenomena, such as causation, knowledge, decay, and the phenomenological feeling of 'moving into the future'. And it is reflected in the use of such phrases as "time's arrow" and in our inclination to say that time "goes" in one direction and not the other. Despite the fact that these expressions have an air of metaphor about them, they clearly imply *anisotropy*—that is, a significant lack of symmetry between the two directions of the temporal continuum. We tend to believe, in short, that time *itself* is temporally asymmetric.

This view of time contrasts with our attitude towards space. We can pick any straight line and define two opposite directions along it. Although the directions are numerically distinct from one another, we would regard them as essentially similar. We wouldn't expect the result of an experiment to depend on the direction in which our apparatus is pointing. Thus we suppose that space is isotropic. Not that this supposition is taken to be *necessarily* true. Aristotelian space, for example, is anisotropic in that directions toward and away from the center of the universe are ascribed quite different causal properties: fire naturally goes one way, and earth another. Similarly it should not be surprising if the question of time's anisotropy proves to be an empirical, contingent matter.

Often, however, those who proclaim the anisotropy of time are not motivated by scientific considerations but are gripped by a certain metaphysical picture. They have in mind that time is more than just a fixed sequence of events ordered by such relations as *later than* and *simultaneous with*, but that it also contains a peculiar property—being *now*—which moves gradually along the array in the direction from past to future. This idea is sometimes combined with a further metaphysical doctrine: namely, that there is an ontological distinction between the past and the future—a distinction that can be represented in a tree model of reality, in which the past consists of a fixed, definite course of events and the future contains nothing but a manifold of branching possibilities. These alleged aspects of time—which I shall describe in more detail as we proceed—are thought to especially distinguish it from space, which possesses no such features. Recent advocates of this sort of view include Broad (1938), Taylor (1965), Gale (1969), Geach (1972), and Schliesinger (1980). On the other hand, there are many philosophers—for example, Russell (1903), Williams (1951), Smart (1955), and Grünbaum (1963)—who reject the 'moving *now*' conception and think that the past and future have exactly the same ontological status. The maintain that the word "now" is an indexical expression (on a par with "here" and "I") whose special function is to designate whatever time happens be the time at which the word-token is uttered. On this account, the thought that an event E is first in the future, will become present, and then fade into the past does not presuppose a 'moving *now*', but it implies merely that E is later than the time at which that thought is entertained, simultaneous with some subsequent time, and earlier than times after that.

Our job in this chapter will be to try to settle these issues—that is, to decide whether there really is any objective feature of the world that corresponds to the idea of a 'moving *now*' and to assess the merits of the tree model. To this end I shall begin by describing and defending McTaggart's (1908) notorious proof that there is no such thing as the 'moving *now*'. But I won't endorse his entire line of thought. McTaggart argues that the 'moving property' theory of *now* is self-contradictory, but he thinks that this conception is nevertheless essential to time. He concludes therefore that time does not exist and that, though "now" indeed functions as an indexical, it refers not to times but rather to other entities that are somewhat like instants of time but only pale substitutes for them. I shall support McTaggart's rejection of the 'moving *now*' but not his further claim that genuine time could not exist without it. We shall see that the best defense against McTaggart's attack on the 'moving *now*' involves a commitment to the tree model of reality. Therefore, in exposing and undermining the antifatalistic and the verificationist motivations for that ontological picture, I hope to reinforce McTaggart's criticism of the 'moving *now*'.

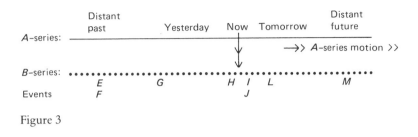

Figure 3

After reaching these conclusions, I shall try to explain why we are nevertheless so captivated by the 'moving *now*' conception. And in the next chapter we shall see that the metaphysical asymmetries suggested by the 'moving *now*' and the tree model are not needed for time to be anisotropic. Even if those ideas are wholly incorrect, there remains the possibility that time is intrinsically asymmetric in virtue of some purely physical, empirical phenomenon.

To begin with, it is worth a moment's digression to note that although McTaggart follows Leibniz (the Leibniz/Clarke correspondance; see Alexander 1956) in trying to prove *a priori* that time does not exist, their two arguments are totally unrelated. This is because Leibniz and McTaggart disagree radically about the sort of thing time would have to be, in order to be real. For Leibniz, real time would be a substance—a Newtonian continuum of thinglike instants at which events are located, ordered by the relation *later than*. But according to McTaggart, something quite different would have to be involved for time to exist: namely, a property, *being now*, which glides along the continuum of instants in the future direction. Moreover there is no need, in his view, for substantial instants. It would suffice if there were merely states of the world ordered by the relation, *later than*, just so long as the property, *now*, moves through these states, singling out progressively later and later ones, as shown in figure 3.

In McTaggart's terminology temporal locations may be specified in terms of two alternative systems of coordinates: the *A*-series, which locates an event relative to *now* (as being in the distant past, the recent past, the present, tomorrow, etc.), and the *B*-series, which locates an event relative to other events (as earlier than *F*, or simultaneous with G, etc.). His view is that time requires that there be a *B*-series, which in turn requires an *A*-series; but that the *A*-series is self-contradictory. Thus Leibniz and McTaggart are arguing against the instantiation of different conceptions of time. Leibniz tries to show that a continuum of instants cannot exist because it would violate the principles of Sufficient Reason and Identity of Indiscernibles. McTaggart contends that the 'moving *now*' model of time is indispensible yet incoherent.

2. McTaggart's argument for the unreality of time

The outline of McTaggart's proof is as follows:

1.   Events are located in a *B*-series (ordered with respect to *later than*), only if time exists.
2.   Time exists, only if there is genuine change.
3.   There is genuine change in the world, only if events are located in a real *A*-series.

THEREFORE:

i.   Events are ordered with respect to *later than*, only if they are located in a real *A*-series.

4.   If events are located in a real *A*-series, then each event acquires the absolute properties *past, now*, and *future*.
5.   There is a contradiction in supposing that any event has any two of these absolute properties.

THEREFORE:

ii.   A real *A*-series cannot exist.

THEREFORE:

(M)   Events are not ordered with respect to *later than*.

Evidently this is a perfectly valid argument: there is nothing wrong with the deductive reasoning by which the preliminary conclusions, i and ii, are derived from their respective premises, and by which McTaggart's final conclusion, (M), is then drawn. It remains, however, to justify these premises. Let us consider what may be said on their behalf.

1. *Events are located in a B-series, only if time exists.*   In order to see that McTaggart's first premise is correct, one must remember that it is not time in the Newtonian sense—an array of thinglike instants— whose reality is in question. Rather, the consequent of (1)—time exists—is supposed to be construed in a very broad way, as something like 'the world exhibits temporality'. And in that case, premise 1 becomes a trivial truth.

2. *Time exists, only if there is genuine change.*   It might seem as though there could be time without change. For consider the scenarios schematized in figure 4. Cases like these are good candidates for

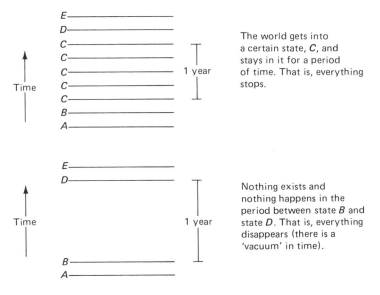

The world gets into
a certain state, *C*, and
stays in it for a period
of time. That is, everything
stops.

Nothing exists and
nothing happens in the
period between state *B* and
state *D*. That is, everything
disappears (there is a
'vacuum' in time).

Figure 4

time without change, and many philosophers who believe there
could be time without change (e.g., Shoemaker 1969) have thought
that it would suffice to show that worlds like those can occur. Such
possibilities, however, are not what McTaggart is intent to deny. His
view is that even in those cases there is still, contrary to first appearances, change of a certain kind taking place: namely, states *A* and *B*
are receding further and further into the past, and *D* is approaching
the present. The *now* is in motion.

According to McTaggart, this sort of change is not only necessarily
present if time passes, but also it is the only sort of *genuine* change that
there could be. Consider, for example, a hot poker, which gradually
cools in the period from *t*1 to *t*2. McTaggart denies that its being hot
at *t*1 and cold at *t*2 constitutes a genuine change. For, he says, it was
and will be true throughout the history of the universe that this poker
is hot at *t*1 and cold at *t*2. Those facts are eternal; they always were,
and always will obtain. That kind of variation with respect to time no
more qualifies as genuine change than a variation of the temperature
along the poker's length. What is required for genuine change, on the
other hand, is that the sum total of facts at one time be not the same as
the sum total of facts at another time.

Here, by the way, is the place at which I would quarrel with Mc-
Taggart's proof, although the rationale for digging in at exactly
this point will become clear only in retrospect. When we see what

he has in mind by "genuine change", this will undermine whatever initial inclination we may have had to agree that the reality of time requires such a thing. In other words, McTaggart's demonstration, in the second part of his argument, that 'genuine change' is self-contradictory should not persuade us that time is unreal but, rather, should force us to acknowledge that time does not require 'genuine change' after all.

3. *There is genuine change in the world, only if events are located in a real A-series.* A variation in the facts would not occur if time consisted in the B-series alone. For the B-series is a fixed ordering of events with respect to one another (and with respect to instants of time, if there are such entities). Therefore the B-series provides only for temporal facts like 'the poker is hot at t1', which, if it obtains at all, obtains forever. Genuine change can come about only in virtue of the relative motion of the A- and the B-series, in which the *now* moves gradually in the direction from earlier to later. This generates genuine changes of the following kind: E is in the distant future, E is in the near future, E is now, E is in the past, and so on.

Note that there are certain metaphysically innocuous construals of the terms "past", "now", and "future" that must be rejected by Mc-Taggart, since they would not imply a real A-series. Consider, for example, the use of "now" in sentences such as "E is now (present) at *t*". This usually means "E occurs at *t*", which is a B-series fact. Similarly "E is past at *t*" means "E is earlier than *t*", and "E is future at *t*" means "E is later than *t*". Past, present, and future have become *relative* properties, whose exemplification is accommodated by the B-series.

Alternatively, suppose that "now" is an indexical expression, like "here" and "I", whose referent depends on the context of utterance. In particular, "now" would rigidly pick out the time, whatever it happens to be, at which the word is used. And suppose that at *t*1 I truthfully say "E is now", and at *t*2 I say "E is not now". Each of these utterances expresses facts, and each of the facts obtains throughout all time. One might be tempted to dispute this claim. One might doubt that "E is now", said at *t*1, expresses a fact that obtains at *t*2, since that sentence uttered at *t*2 would be false. But this would be a non sequitur because the sentence does not say the same thing at the two different times. The word "now", used at *t*1, simply provides a way of referring to the time *t*1. And the fact expressed by the first remark—though perhaps not the same as the fact expressed by "E is at *t*1"—is just as permanent. Consequently McTaggart holds that for there to be genuine change and a real *A*-series, "past", "present", and "future" can be neither relational predicates nor indexicals.

So far McTaggart has tried to show that time requires the existence of a genuinely moving *now*. And, as I have already said, this preliminary conclusion may be resisted. The remainder of his argument is a demonstration that the 'moving *now*' conception is self-contradictory. This is part of his reasoning that I believe is correct and important.

4. *If events are located in a real* A-*series, then each event acquires the absolute properties past, now, and future.*   A real A-series entails that for every event such as *E*, there is a fact, included in the totality of facts that constitutes the universe, consisting of *E*'s having the quality of *presentness*, that is,

E is   (or, E is now)

but also the universe must contain the facts

E will be   (or, E is future)

and

E was   (or, E is past)

Given what is meant by "a real *A*-series," such facts are not relations between events and times. They are not, in other words, the exemplification of merely *relative* properties, which can both apply and fail to apply to the same event relative to different frames of reference. Rather, such facts consist in the exemplification by events of absolute properties.

5. *There is a contradiction in supposing that any event possesses any two of these absolute properties.*   Past, present, and future (which are equivalent to 'earlier than now', 'now', and 'later than now') are incompatible attributes. Therefore the supposition that one event has them all involves a contradiction. That is to say, it is impossible that the history of the universe contain the three facts: *E* is past, *E* is now, *E* is future.

One will be tempted to object, as follows. There is a contradiction only if the *A*-series qualities are attributed *simultaneously* to *E*; but such simultaneous attribution is not required by the existence of the *A*-series; rather, its existence entails only that each of the *A*-series qualities apply to *E* at some time or other. That is to say, McTaggart's premise 4 will be satisfied even if the *A*-series determinations are acquired *successively*, and in that case no contradiction arises. In other words, the requirement described in premise 4 may be met by the existence of the facts

> E is future at $t1$
> E is present at $t2$
> E is past at $t3$

which are quite compatible. There is no need to take premise 4 to imply that all the A-series determinations would have to apply at the same time.

However, one must beware of resolving the contradiction in ways that involve eliminating any real A-series. And this is exactly what has just happened. For the meanings of "future", "present", and "past" in the preceding sentences are "later than", "simultaneous with", and "earlier than". The facts described are generated by the B-series. Genuine change has been lost in the reformulation. To preserve genuine change—to have a real A-series—it is not enough that there be a variation in *relative* presentness from one time to another (like the variation in the velocity of an object relative to different reference frames). Rather, there must be variation of facts. Thus it is necessary to construe premise 4 in such a way that the transitions from 'E will be' to 'E is' to 'E was' are transitions between mutually exclusive, absolute states.

At this point McTaggart's opponent might well complain that revealing such a variation of facts was precisely the intention behind his reformulation of premise 4. The idea, he says, was *not* to transform *past, present,* and *future* into mere relations (which admittedly only succeeds in eliminating the A-series) but rather to suggest that the facts "E is past", and so on, might themselves obtain only relative to a temporal perspective. In other words, the premise 4 should have been formulated more perspicuously with the following sentences:

> The fact that E *is future* obtains at $t1$
> The fact that E *is present* obtains at $t2$
> The fact that E *is past* obtains at $t3$

Thus there *is*, after all, a variation, from one time to another, as to which facts obtain.

In response to this suggestion, however, we are justified in resisting the crucial assumption that the italicized internal sentences express facts. For a strong case can be made that this latest formulation of premise 4 trades on an idiosyncratic and unmotivated conception of *fact*. After all, we do not regard

X is to the left of Y

and

X is not to the left of Y

as explicit descriptions of facts. Rather, we suppose that whenever such claims are true, they are partial accounts of facts whose explicit descriptions take the form

X is to the left of Y relative to Z

and

X is not to the left of Y relative to W

Similarly one does not say that the facts, fully articulated, include

It is raining

and

It is not raining

But rather, for example,

It is raining in Manchester

and

It is not raining in Florida

The general point is that we reserve the term "fact" for those aspects of reality whose explicit descriptions are sentences that are true *simpliciter*—and not merely true relative to some context or point of view, and false relative to others. Consequently, if we are going to say that "E is past" is sometimes true and sometimes false, then unless some good reason is given to depart from our usual conception of fact, we should not countenance this sentence as an explicit characterization of a fact. The real facts, as we said initially, are described by sentences of the form "E is past at t", in which pastness has been transformed into a relation.

These remarks do not absolutely preclude the idea that facts may be relative: that is, dependent on a frame of reference. The point is, rather, that such a perspectival view of reality would require a radical change in our conception of fact, and that any such revision would call for some independent motivation. So far, in our discussion of this problem, no reason to abandon the usual notion of fact has been offered. And this is why the response to McTaggart that we are now considering is inadequate as it stands. However, that is not to say that no such argument for perspectivalism *could* be given. Indeed, a strategy to that end, based on verificationist considerations, is suggested by Dummett (1960). I shall take it up in the next section, in connection with Aristotle's tree model of reality.

I have been arguing that McTaggart's contradiction is not avoided by the supposition that the futurity, presentness, and pastness of E obtain relative to three times, $t1$, $t2$, and $t3$. Notice that it is equally futile to try to escape his conclusion by rendering the facts as follows:

> E is future, in the past
> E is now, in the present
> E is past, in the future

In the first place, this strategy is subject to the same criticism as before: the initial occurrences of "future", "present", and "past" have been transformed into relative properties. So these sentences can be reformulated as

> E is later than past times
> E is simultaneous with the present time
> E is earlier than future times

which do not entail the existence of the facts required by a real *A*-series. And in the second place, such second-order temporal attributions are just as problematic, from McTaggart's point of view, as the first-order ones. For they are compatible with one another only if we assume that the *past, present,* and *future* are disjoint regions of time (or of events). And that assumption is contrary to his requirement: that every event and time has the qualities of *past, present,* and *future*. This being so, we can derive from the first statement (supposing that "past" and "present" are coextensive)

> E is future, in the present

which conflicts with the second statement. Therefore the contradiction is not avoided by introducing second-order temporal attributions. This is because, from the fact that each of the first-order attributions must hold, it follows that each of the second-order attributions must hold. And they conflict just as blatantly as the first-order attributions.

The most common criticism of McTaggart's argument (e.g., Broad 1938; Prior 1967) is exactly the point just dealt with: to claim that consistency may be achieved by a reformulation in terms of higher-order temporal attributions. It is not appreciated that McTaggart himself considers and refutes this strategy. To repeat, he denies that his requirement that the world contain the facts

E is past
E is present
E is future

is misstated when construed literally, in which case the facts are mutually inconsistent with one another; and therefore he denies that the required facts are accurately represented by, for example,

E is past, in the future
E is now, in the present
E is future, in the past

For the operative occurrences of "past", "present", and "future" have been turned into relations. Therefore McTaggart denies that the initial contradiction is treated by introducing second-order attributions. Nevertheless he is quite happy to conduct the argument at the second level. For, from his first-order requirement, it follows that *every* second-order attribution must hold—and this is also a contradiction.

Thus McTaggart shows that a certain very tempting, 'moving *now*' conception of time is not actualized. But he does not succeed in proving that time is unreal, because the first part of his argument is not persuasive (Mellor 1981). In other words, we need not agree with him (premise 2) that it is essential to the reality of time that there be 'genuine change', in his sense. This claim is implausible and never really substantiated. If we are persuaded, as I think we should be, by the second part of his argument, we will conclude that there can be no 'real *A*-series' or 'genuine change'. Rather, change is always variation in one thing with respect to another, the totality of absolute facts about those functional relations remaining forever constant.

3. The tree model of reality

Affiliated with the 'moving *now*' conception of time is another unorthodox metaphysical theory—roughly speaking, that only past and present events exist, and future ones do not. Reality, in this view, is thought to resemble an unlimited tree: from any point, there is a single definite path downward (history is fixed) but above each point we encounter a proliferation of many possible branches (the future is open). Thus statements about the past and present are, right now, determinately true or false, unlike current claims about the future, which do not attain a truth value until the predicted events either occur or fail to occur. Only the advance of the *now* settles which path through the tree is taken and which predictions are true.

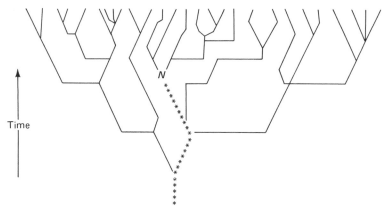

Figure 5

In figure 5, *N* represents the present state of the world. The chain of stars represents the fixed past. (Note that there is just one way down the tree from N.) And the branches growing up from *N* represent the many things that might happen later; thus the future is open. The only statements about the future that hold at *N* are those that are obtain in all of the branches that stem from *N*. If something—for example, a sea battle tomorrow—occurs in some branches but not in others, then from the perspective of *N*, there is no fact of the matter as to whether there will be a sea battle tomorrow.

Aristotle, with whom this sort of view is often credited, was not the first philosopher to deal with time; but he was the first to offer more than provocative aphorisms and to try, in a scientific spirit, to clarify and demystify our conception of temporality. And he reached the conclusion that time is *doubly* asymmetrical. In the *Physics* (Book IV) he endorses the 'moving *now*', and in *De Interpretatione*—according to one natural construal of it—he advocates a truth-value asymmetry as the only way to avoid fatalism.

I am going to argue, on the contrary, that there is no ontological asymmetry between past and future and that the threat of fatalism can be averted without radical measures of this sort. To begin with, however, I want to explore the relationship between the tree model of reality and the 'moving *now*' conception of time. We shall see that although advocates of the 'moving *now*' can base a defense against McTaggart on the tree model, there is, on the other hand, no incentive for advocates of the tree model to endorse the 'moving *now*' conception of time.

On the latter point, notice that the tree model does not preclude an indexical construal of "now"; and so there is no compulsion to hold that reality contains not merely the tree but also a quality—*now*—that moves up the tree, selecting which branches are to be actual. To see more vividly why this addition to the tree model is not needed, suppose that the past, as well as the future, were not determined by the present. Suppose, in other words, that more than one possible course of history were compatible with the current state of the world. Then a network of possibilities would branch out into the past. And in that case both the past and the future would be open. One would think that if the openness of the future called for a future-directed *now*, then the openness of the past would similarly entail a past-directed *now*. Yet there surely would be no philosopher who would go quite so far as to postulate a *pair* of oppositely moving *nows*. (When would they meet?) This suggests that a fondness for the tree model does not produce a taste for the 'moving *now*' conception of time.

On the other hand, one can quite understand why advocates of a 'moving *now*' would be inclined to sympathize with the tree model. McTaggart's argument that the 'moving *now*' does not exist depends on exposing a contradiction between facts "E is past", "E is present", and "E is future", all of which must, given a genuinely moving *now*, belong to the totality of absolute facts in the world. However, as Dummett (1960) has observed, this argument requires the assumption (which, as we shall see in a minute, is questionable) that there *is* such a totality of facts. If there is no such thing—if the facts change from one temporal perspective to another—then the only troublesome contradictions are contradictions from a particular temporal perspective. But a 'moving *now*' does not require that E be past, present, and future from a single temporal perspective. So if there is no time-neutral body of absolute facts, there is no contradiction. Thus, by denying the assumption of this totality, McTaggart's objection can be sidestepped.

But only at substantial cost. For the crucial move—denying the assumption that there is a totality of facts—seems quite bizarre, unless it is independently motivated. As we saw in our discussion of McTaggart's proof, it looks simply ad hoc, and contrary to our usual conception of fact, to say

> The fact that E *is future* obtains at $t1$

rather than

> E is the future at $t1$

or, in other words

> E is later than $t1$

Thus there might seem to be no reason to countenance facts that obtain at some times and not at others. The attraction of the tree model of reality is precisely its ability to supply this rationale. For the tree model purports to show, independently of anything to do with *now*, that there is no complete, time-neutral body of facts. At any point the facts consist of a certain course of history, plus the present state and whatever is in every future branch. But this body of facts changes. And there is no summing up, from a temporally neutral point of view, to obtain an overall picture of reality. Thus defenders of the 'moving *now*' will be happy to embrace the tree model. For, in the context of that model, their best reply to McTaggart is not ad hoc: genuine change can be achieved without contradiction. By the same token a thorough criticism of the 'moving *now*' conception must eventually deal with the tree model of reality. Let us therefore examine the reason that, following Aristotle, is most often cited as motivating this ontological asymmetry: the avoidance of fatalism.

4. Fatalism

The case for fatalism goes something like this. What was true in the past logically determines what will be true in the future; therefore, since the past is over and done with and beyond our control, the future must also be beyond our control; consequently, there is no point in worrying, planning, and taking pains to influence what will happen.

The fatalist's assumption that the future is already logically determined derives from his supposition that if some sentence, such as "A sea battle is occurring", would be true if asserted at a future time, say, on January 1, 1999, then it would always have been correct to predict that a sea battle will occur on that date. In particular, on January 1, 1000, it was the case that a sea battle will occur on January 1, 1999. But this past fact—namely, "On January 1, 1000, it was the case that there will be a sea battle on January 1, 1999"—logically entails "A sea battle will occur on January 1, 1999". This is the rationale for the fatalist's assumption of logical determinism of the future by the past. Aristotle's idea is to avoid fatalism by rejecting this deterministic thesis. He undermines its rationale by maintaining that contingent statements about the future are presently neither true nor false.

More precisely, the argument for fatalism proceeds as follows. Let f refer to an arbitrary future time and p to an arbitrary past time; let capital letters stand for statements that are about free actions, and $\Box$ . . . stand for "It is now determined, and beyond our ability to influence the fact, that . . ."; and let, for example, Qp mean "Q was true at past time p", and Sfp mean "It was the case at past time p that S will be true at future time f".

We begin with a premise intended to express the idea that due to the future orientation of causation, the past is fixed and beyond our control:

$$Qp \rightarrow \Box(Qp) \tag{1}$$

Then we substitute for Q the statement, "S will be true at future time f" and obtain

$$Sfp \rightarrow \Box(Sfp) \tag{2}$$

But Sfp necessarily has the same truth value as Sf, and if one of these facts is beyond our control, then so is the other. Therefore we can replace the antecedent of (2) with Sf, and its consequent with $\Box(Sf)$, to get

$$Sf \rightarrow \Box(Sf) \tag{3}$$

If we can assume that there is now a definite fact about whether some act, described by R, will or will not occur in the future, then we can say

$$Rf \vee (-R)f \tag{4}$$

But now we may substitute, first R, and then $-R$, for S, and obtain

$$\Box(Rf) \vee ((-R)f) \tag{5}$$

which says that either R's future truth is now fixed and beyond our control, or R's future falsity is fixed and beyond our control.

It seems clear that the reasoning involved in this argument is perfectly valid. And if so, then our options are as follows:

i.   Abandon the plausible sounding view that the past is already determined and beyond our control—that is, deny (1).

ii. Follow Aristotle by giving up the idea that all statements about the future have a present truth value—that is, deny (4).

iii. Invite fatalism by agreeing that even future events are presently beyond our control—that is, accept (5).

None of these alternatives looks attractive. However, some are clearly worse then others. In particular, the practical consequences of fatalism make that doctrine literally impossible to accept, and therefore make it impossible for us to accept statement (5). Moreover, the

Aristotelian renunciation of facts about the future runs counter to logical principles (e.g., that every statement is either true or false) and a conception of time that are extremely plausible and have been pillars of science and common sense. So we would very much like to avoid options ii and iii. Why then resist the first option? Why not agree that, in some sense, the past is not beyond our control? Is it so paradoxical to allow, for example, that I can now decide whether or not it was true last week that I would scratch my head today?

No sooner is it formulated, than this escape from the paradox seems obviously right. But why wasn't this clear from the beginning? Why did we find premise (1) so plausible? The answer is that there is a definite sense in which the past *is* beyond our control. We would doubt the sanity of anyone who announced his intention to do something now in order to bring about some past *event*. So, at first sight, there is a conflict between our way out of the choice between fatalism or Aristotelianism and our respect for the fact that effects do not precede their causes.

But of course there really is no such conflict. The way to reconcile our influence over the past with the direction of causation is to recognize that S may be true at time p without there being any concrete event or state of affairs at p that makes S true at that time. It is a failure to notice this simple point that is responsible for the paradox. Consider:

The wheel was invented several thousand years ago in Egypt

Suppose this is true. There need be nothing *now* (or *here*, for that matter)—no present occurrences—that make it true. What makes it true, if it is true, are certain events that took place in Egypt thousands of years ago. Similarly, even though it was true last week that I would now scratch my head, there was no event or state of affairs last week that made it true.

Therefore, although my present decision does, in some sense, influence the past—since it was responsible for making it true in the past that I would scratch my head—my decision did not bring about any past event, and so there is no conflict with the principle that events never precede their causes. Therefore we can happily deny premise (1) of the fatalist's reasoning. Thus Aristotle's argument for an ontological time asymmetry is unconvincing. It derives, I have suggested, from mistakenly supposing that premise (1) is justified by the direction of causation—a mistake that derives in turn from a confounding of facts with events, or, more accurately, from confusing a proposition's being true at a given time with the existence of concrete circumstances at that time in virtue of which the proposition is true.

5. Verificationist roots of the tree model

But is this conclusion really fair to Aristotle and his sympathizers? Perhaps he was not at all confused on this point? Perhaps one might maintain, quite deliberately, that a fact obtains at a time only in virtue of something going on at that time? If so, then premise (1) will be justified, the tree model will begin to look much more attractive, and one will be in a position to defend the 'moving *now*' against McTaggart's objection, as we saw earlier, by denying the existence of an absolute totality of facts. Let us explore this possibility.

Aristotle seems to think that the state of the world at any given time contains determinants of every past event but does not determine everything (e.g., free actions) that will happen later. In other words, he holds that the relation of physical determination is time-asymmetric. And, as we shall see in chapter 3, such an asymmetry will suggest, perhaps even entail, that time itself is asymmetric (or anisotropic). However, there is no immediate implication that future events do not enjoy exactly the same ontological status as past events—namely, the status of existing at some moment of time or other. One can, it seems, endorse a tree model of *possibility*, thereby avoiding fatalism, without being committed to a tree model of *actuality*.

Thus the tree diagrammed in figure 5 may be taken to represent an Aristotelian world in which relations of physical possibility are asymmetrical in time. The laws of nature in this world are indeterministic in such a manner that given the state of things at any time, there is only one course of history that could have led up to that state but there are many alternative ways that the future could go. However, unlike Aristotle, it would seem that one might perfectly well claim that only one of these ways will be actual and that future reality (which could be represented by a continuing chain of stars) is quite definite. From the point of view of the present state of affairs, N, the future is not determined and not predictable, as the past is—but it is no less real.

Aristotle's position, however, is that no particular future path is singled out as actual—there should be no stars above the N in figure 5. He can agree there are *some* truths about the future. It will be the case that either a sea battle takes place tomorrow or it doesn't; for that disjunction is true in every branch. But he must deny that the battle either will or will not take place; for a battle occurs tomorrow in some branches but not in others, and no particular branch is especially relevant to the question of what the future holds.

It will be conceded by proponents of the Aristotelian position that most of us implicitly reject their view. We might be persuaded that the avoidence of fatalism calls for a tree model of possibility. But we do not suppose that a fact may obtain at a time, only if there are concrete events at that time to make it hold. Therefore we see no reason to go so far as a tree model of actuality. On the contrary, most of us do believe that some definite course of future events will occur whether determined or not. We naively make a distinction between what *will* happen and what *must* happen. And such a distinction is recognized in practical affairs when we give credit (e.g., by paying off bets) for correct predictions. So the question arises: What could possibly count against our ordinary way of thinking, and in favor of the counterintuitive Aristotelian position?

Antirealist theses are typically motivated by verificationism—by the thought that we can understand a sentence only to the extent that we know how to recognize if it is true (Dummett 1978). This is the route to intuitionism in mathematics and behaviorism in psychology, and the domain at issue here is no exception. According to our antirealist regarding the future, we now understand a prediction that some event will occur tomorrow only if we are now able to tell whether such a prediction is correct or not. And we can do this only if there exist present determinants of the truth, or the falsity, of the prediction. But usually there are no such determinants. So, in those cases, there is no sense to the idea that the prediction is true. Therefore, claims the antirealist, the elements of our linguistic practice that presuppose a definite future are incoherent and unintelligible. We are mistaken in thinking that we truly understand such talk of the future.

Thus extraordinarily strict standards of intelligibility are presupposed by the antirealist; and the plausibility of his tree model hangs precisely on the question of whether those standards should in fact be adopted. Can we be satisfied with ordinary understanding as revealed by a facility with a stable, useful, linguistic practice? Or should we insist on the sort of super '*understanding*' provided by strict verificationist standards? I myself am not moved by a desire to streamline language along verificationist lines and, in particular, see nothing unfortunate or ontologically significant in the fact that statements about the future are not often verifiable in advance. But clearly a proper discussion of verificationism would take us too far afield. In its absence we can perhaps make do with the following uncontentious conclusion. '*Reality*'—defined in terms of what we '*understand*'—conforms to the tree model; but common or garden reality, described by means of what we ordinarily take ourselves to understand, does not. In other words, let us grant the verificationist his special, sanitized conceptions of *meaning, truth,* and so on; allow that certain instances of "*p* or not *p*" are *false*, in his sense; but proceed to operate with our own convenient notions, without regard for the standards of verifiability that they fail to meet. Insofar as this attitude is legitimate, we will have no use for the tree model and, therefore, no perspective from which to rescue the 'moving *now*' from McTaggart's refutation.

6. Our sense of passage

If the 'moving *now*' conception of time is wrong, then why is it so compelling? A neat answer to this question (Grünbaum 1973) is that because the word "now" functions somewhat like a noun, we are seduced into regarding it as standing for a single entity whose varying locations are instants of time. But the direction of motion of any object is, by definition, the set of its locations ordered according to the times at which it possesses them. In this way we evolve a picture of a mysterious entity, *now*, gliding inexorably into the future.

Thus the asymmetry involved in regarding the future, rather than the past, as *the* direction of time, seems to stem entirely from the time bias built into the meaning we have given to the expression "*the* direction of change" and does not arise from any asymmetry within the temporal continuum itself. Indeed, our temptation to recognize such a 'substantial' asymmetry gives every indication of being a good example of what Wittgenstein constantly warned against: the derivation of misleading metaphysical conceptions from structures in language:

> In our failure to understand the use of a word we take it as the expression of a queer *process*. (As we think of time as a queer medium, of the mind as a queer kind of being.) (1953, sec. 196)

However, one cannot be fully satisfied with this sort of answer until the '*feeling*' of time flow has been explained. There appears to be something about the structure of our awareness of things that makes it very natural for us to characterize ourselves as "moving into the future", or as "perceiving the passage of events as if they were floating on a river that flows past us". But what are the aspects of experience that produce our sense that time passes? And how do these phenomenological features contribute to the attractiveness of the 'moving *now*' conception?

Let us consider, therefore, the temporal structure of a typical experience. (The following account is derived from Izchak Miller's (1984) elucidation of Husserl's *Phenomenology of Internal Time-Consciousness* (1928)). In the first place, it is significant that a normal

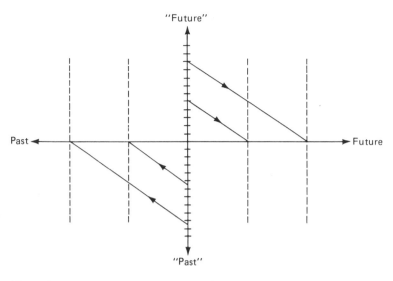

Figure 6

experience is a complex entity made up of, among other things, memories of the distant past, more recent recollections, sensations, and anticipations projected for various times in the future. Moreover some of the memories and anticipations are of experiences that themselves contain just those sorts of constituents. Thus, each normal experience involves not only an awareness of variation in the physical world but also an awareness of the fact that there has been, and will be, a sequence of experiences in which what is anticipated will be sensed and subsequently remembered. In other words, an experience represents both a set of events strung out in time and a set of experiences, each of which represents, from different perspectives, the same string of events.

In figure 6 the central vertical line stands for my experience at a given time, and the rest of the diagram pictures part of the intensional content of that experience. The points in the bottom half of the line are memories, the central point is a sensation, and the upper points are anticipations. The horizontal line stands for the stream of events in time that are represented by these various components of my experience. The intensional content of each component is indicated by the arrow leading from it. Thus each component (memory, sensation, or anticipation) attributes a particular distance from *now* (respectively, negative, zero or positive) to a specific event and also attributes experiences (shown by vertical dashed lines) to those times. The components of these experiences bear a systematic relationship to the experience I am now having. Thus I anticipate that things I now re-member will at some later time be remembered as having occurred even further in the past, that the things I now sense will be remem-bered, and that some things I now anticipate will then be expected to happen not so far in the future. And I remember that many of the things I now remember were once remembered as not having happened so far in the past, that the things I now sense were once anticipated, and that things I now anticipate were then expected to happen even further in the future than they now are.

In virtue of the phenomenological difference between memory and anticipation, this picture of our experience is asymmetrical. It is therefore able to accommodate certain elements in our conception of time, such as the phenomenological difference between past and future orientations. But it remains unclear why there should be a sense of 'movement' through time.

Some philosophers have been inclined to attribute this feeling of time flow to the fact that memories are taken to be more reliable than anticipations, so that we are conscious of a continual accumulation of perceptual knowledge. But this does not help to account for a sense of motion (Smart 1980). In the first place, it is left unexplained why we should seem to move through time in the direction of *increasing* knowledge, rather than in the opposition direction of *decreasing* knowledge. And second, one can imagine there being a limit to mem-ory, so that every experience is automatically forgotten after, say, five seconds (as in the case of those who suffer from Korsakoff's syn-drome). In such circumstances there would be no change in our total quantity of perceptual knowledge and yet we would still experience the passage of time.

What then is capable of transforming our awareness of the variation in experience across time into a sense of time flow? I think that the answer is suggested in what I said initially about the so-called motion of *now*. That is, all we mean by "the direction of change" of any process is the direction *from* the earlier states of the process *to* the later states. In particular, the direction of motion of an object in space is, by definition, *from* its earlier locations *toward* its later locations. Now, as I have just described, we are aware of a succession of com-plex experiences. Each has the *same* structure consisting of a present sensation, anticipations with various degrees of projected futurity, and recollections of various types. And each has roughly the same content—a set of phenomena strung out in time. The difference be-tween them is that the later that an experience is represented as occur-ring, the more pastness and less futurity it attributes to any given event. Thus we are conscious of the same experiential framework being filled with the same contents from different temporal perspectives. Therefore it seems to us as if a single entity—the structure of experience—is undergoing these changes. However, as I have been emphasizing, the direction of a change is *from* its earlier *to* its later states. Thus we obtain a sense of our consciousness in motion relative to the world—either crawling into the future or else interacting briefly with events as they rush into the past.

In summary, I explain our 'sense of the passage of time' as the product of two factors: phenomenological and linguistic. In the first place, the framework of a single experience contains elements of anticipation and recollection that picture time as a one-dimensional continuum, distinguish the two directions along that continuum, locate a string of physical and mental events in it, and also locate an array of various states of awareness of those same events, each from a different vantage point. This is the substantive phenomenological component in our awareness of time flow. In the second place, our conventions concerning the concepts of 'motion' and 'direction' lead to a particular way of describing the array of states of mind that is pictured in any single experience. Since each state consists in our experiential framework, located at a given phenomenological time and filled with the content that is appropriate for that vantage point on the string of events, the change of state that we are aware of is a variation in the temporal *location* of our experiential framework. Thus we describe this change as "movement through time". And since, by convention, we suppose that *the* direction of any process is *from* earlier *to* later stages in the process, so we come to say that our experience involves a sense of motion into the future.

3

Anisotropy

1. The meaning of "anisotropy"

The fact that *now* coincides with progressively later and later events indicates nothing about the structure of time. In particular, it does not imply that time has an 'arrow' or a directional character of any sort. For, as we have seen, "now" is an indexical expression, like "here", so there is no metaphysical significance in the variable location of its referent. However, there remains the possibility that time is intrinsically asymmetric. It still could be that time exemplifies some asymmetry, having little to do with the behavior of *now* but somehow related to patterns of events that are asymmetric with respect to time. Perhaps, in other words, the glaring time biases displayed by phenomena such as knowledge, causation, and dissipation of order point the way toward an asymmetry within the structure of time itself. It could be that the two opposite directions along the temporal dimension are significantly different from one another, even though neither has the metaphysically special status of being the direction "in which time goes". Thus we should be open to the idea that time is anisotropic, despite having no *privileged* direction.

In order to assess this possibilty, we must start by explaining what precisely is meant by the supposition that time is anisotropic. Unfortunately, few writers on the subject of temporal anisotropy provide an explicit account of that notion. Most either assume that the general idea is already clear enough, or they offer a definition involving terms that are equally obscure. They proceed to endorse one or another criterion of temporal anisotropy; but they are unable to justify any such choice in the absence of a satisfactory understanding of what their criterion is supposed to identify.

I shall devote the present chapter to this problem. I intend to argue that the anisotropy of time would consist in an *intrinsic* dissimilarity of the past and future directions: a difference that would be manifested in some time asymmetry within our laws of nature. To begin with, I'll explain this idea. Then I'll support it in the face of competing suggestions by Black, Grünbaum, and Earman. And, finally, I'll argue that the current empirical evidence indicates that time itself is intrinsically symmetric. The issues under discussion here are self-contained and play only a minor role in what follows. Therefore the reader, especially one who is sympathetic to my conclusions, could skip or skim this chapter without missing much that will be needed later.

A relatively explicit, yet still far from adequate, characterization of anisotropy is given by Reichenbach (1956, pp. 26–27):

> When we say that a line, though serially ordered, does not have a direction, we mean that there is no way of distinguishing structurally between right and left, between the relation and its converse. In order to say which direction we wish to call "left," we have to point to the diagram; or we may give names to points and indicate the selected direction by the use of names. Had we decided to call "right" what we called "left," and vice versa, we would not notice any structural difference; that is the relation *to the left of* has the same structural properties as the relation *to the right of*.
>
> It is different in the case of the linear continuum of negative and positive real numbers, which we can map on a straight line. The numbers are governed by the relation *smaller than*, which is asymmetrical, connected, and transitive, like the relation *to the left of*, therefore the numbers have an order. But in addition, the relation *smaller than* has a direction; that is, it is structurally different from its converse, the relation *larger than*. This is seen from the fact that we can distinguish between negative and positive numbers in the following way. The square of a positive number is positive, and the square of a negative number is also positive. We therefore can make this statement for the class of real numbers. Any number which is the square of another number is larger than any number which is not the square of another number. We thus have defined the relation *larger than*, and with it, the relation *smaller than* in a structural way.
>
> Applying these results to the problem of time, we find that time is usually conceived as having not only an order, but a direction. The relation *earlier than* is regarded as being of the same kind as the relation *smaller than*, and as not being undirected like the relation *to the left of*. This means that we believe that the relation *earlier than* differs structurally from its converse, the relation *later than*.

Despite his use of the expression "having a direction", with its potentially misleading 'moving *now*' connotations, Reichenbach is in fact concerned with the general concept of anisotropy. The unsatisfying aspect of his account is that anisotropy (or, in his terms, directionality) is defined as the existence of a *structural* difference between an ordering relation and its converse; but he gives no explanation of the notion of structural difference. Instead, we are presented with two examples: a line in space and the series of real numbers, and we are told some features of these examples that render them, respectively, isotropic and anisotropic. It is left to the reader to discover the right way of generalizing from these illustrations and, in particular, to work out what anisotropy (or structural difference) will amount to in the case of time.

Can we, therefore, extract from Reichenbach's examples, and other compelling intuitions, a clear characterization of anisotropy? We seek an account that will explain why it is that when an array is anisotropic, its two directions may be distinguished solely in terms of the properties of the elements; and without pointing or giving names to particular elements (as would be required along a line in space). In addition the account should help us to understand why the existence of time-asymmetric laws of nature is generally taken to guarantee time's anisotropy.

To get a more vivid sense of what we are looking for, imagine an endless, straight tube. If it gradually becomes narrower in one direction, then we would regard the tube as anisotropic. But if it is a prefect cylinder and everywhere a constant color, temperature, etc.—the same in both directions—then the thing would seem isotropic, and a direction along it could be specified only by reference to particular parts of the tube.

Similarly, when we say that time is isotropic what we mean is that the two directions are in some sense the *same*. Not of course that they are strictly identical. There are, after all, *two* directions. But "the same" in whatever sense we employ those words when we say that two regions of homogeneous space, or two electrons, are the same. What sense is this? Presumably it implies an exact similarity in certain respects—the sharing of properties of a certain sort. But what sort? Not every property need be shared. Two electrons don't have the same position or the same constitutents. Two regions don't contain the same objects; nor are they located at the same distances from other parts of space. The usual answer, which I think is good enough for our purposes, is that two electrons or two similarly shaped regions of space are *intrinsically* (qualitatively) identical—sharing all their "intrinsic properties": where the intrinsic properties of an object are, roughly speaking, those that do not involve relations to anything other than the object itself or its parts. Some intuitive examples of intrinsic properties are *being an electron, being spherical, being made of plastic*, and some examples of non-intrinsic (also known as relational) properties are *being one mile from the earth's surface, being soluble in water*, and *liking bananas*. Now, what the difference is, in more precise terms, between intrinsic and relational properties is an old and hard problem whose proper examination would take us far from our present concern. (See Bernald Katz 1983 for discussion of some of the difficulties). For our purposes it is enough that we be sure that such a distinction exists, and that we have some confidence in how to deploy it in particular cases. To that end, let me go into a little more detail about it.

We can often identify relational properties from the predicates that are used to express them: they contain names ("is twenty-four hours from Tulsa") or unrestricted quantification ("orbits some star"). However, this way of detecting relational properties only works under certain conditions. In particular, it requires that the predicates be composed from a limited vocabulary of so-called "natural" predicates: namely, predicates that play a role in articulating laws of nature. Again, notorious difficulties arise in connection with this idea that we can't attend to here (see Quine 1969; Shoemaker 1980; Lewis 1983). The important point is that if the restriction to composition from 'natural' predicates is ignored, then our test for relational properties will fail miserably. For there will be nothing to stop the use of contrived predicates like "grue" (meaning "green and examined before time T, and blue otherwise") and "bleen" (meaning "blue and examined before T, and green otherwise") in the construction of complex predicates that will disguise the true intrinsic nature of the properties they express. For example, it could be argued that *being green* is relational, on the ground that it is expressed by the predicate "is grue and examined before time T, or bleen and examined after T" (Goodman 1955). Thus we should think of the intrinsic properties of an object as those expressible by predicates that are composed of natural predicates, contain no names, and have no quantifiers except those restricted to range over just the object itself and its parts.

Given this notion, we can say that the isotropy of time would consist in the intrinsic identity of its two directions: their having the same intrinsic properties. But how can one tell if this is so? Which features of a temporal direction are intrinsic and which relational? Let me clarify our criterion by examining the implications for isotropy of three kinds of property—particular, general, and nomological—that might be possessed by a direction of time. First, there are *particular* properties, such as those we implicitly attribute to the future direction when we give the time order of two particular events:

> *E* is later than *F*

or

> The future direction goes from *F* to *E*

The characteristic expressed here is not shared by the past direction. This disparity does not imply that time is anisotropic, however. For not only is the property obviously relational (involving specific events, *E* and *F*), but its happening to apply to one direction and not the other does not suggest any intrinsic dissimilarity between them. Second, consider the *general* property that we attribute to a direction of time when we say that *all* events of a certain kind are followed by events of a certain other kind. Again, we have a difference between the past and future directions. Again, however, this difference does not necessarily imply an intrinsic dissimilarity between them. For the existence of this kind of disparity might be explained in terms of the initial conditions of the universe—that is, by facts having nothing to do with time itself—so it need not provide evidence for any intrinsic difference between the past and future directions. Finally, there are the *nomological* features that would be ascribed to a direction of time by a time-asymmetric laws of nature: for example,

> As time goes on, the entropy of an isolated system never decreases

Since there is no analogous law regarding the opposite direction, we have here a property that applies as a matter of physical necessity to one direction and not the other. The possession of such a property does not *constitute* an intrinsic difference between the directions, since it mentions physical systems and is therefore relational. However, we do have grounds for *inferring* an intrinsic difference between the directions. In supposing that it is by *law* that one of time's directions can coincide with an entropy increase and the other cannot, we are excluding the possibility that this difference depends on something other than time—say, the initial conditions of the universe. Thus we are supposing that there must be something about time *itself* that explains the difference. Thus a sufficient condition for there to be an intrinsic dissimilarity between the past and future directions of time is that they be distinguished by laws of nature. And this will be manifested in some difference between the ways in which *earlier* and *later* function in the laws of nature. In other words, time is anisotropic if, for some law involving the concept *later than* in some way, the result of replacing that concept with *earlier than* is not a law.

I have been arguing that time-asymmetric laws of nature are a *sufficient* condition for time to be anisotropic. But there is no reason to regard this condition as *necessary* for anisotropy. For we cannot preclude the possibility of (future) physical theories in which some of time's intrinsic features will be treated as *de facto*, that is, as not required by law. In that case there could be intrinsic differences between the past and future directions that are not reflected in any time asymmetry in laws of nature. Therefore, the isotropy of time is not absolutely guaranteed by the time symmetry of laws. However, although the conceivability of such de facto intrinsic differences must be acknowledged and might one day become a live prospect, this idea will play no role in what follows. For our present understanding of time is not deep enough to give even the faintest hint as to what such differences might be like, or how they could arise from de facto features of the world. At the moment the most likely form of evidence for anisotropy would be the discovery of time-asymmetric laws.

In the following sections I will sharpen the conception of anisotropy that I have been presenting, by contrasting it with some alternative proposals. I will examine attempts to define anisotropy in terms of the objectivity of time order (section 2), de facto asymmetries, (section 3), and nomologically irreversible processes, (section 5); and I will defend the idea that the "directional character" of time might vary from one region to another. Finally, having gained a clearer sense of what the issue precisely amounts to, I will argue that time is in fact isotropic: its two directions are intrinsically the same.

2. *The objectivity of time order*

I have just proposed that anisotropy be explained as an intrinsic difference between the two directions of time and that this would be manifested most clearly by a time bias in the laws of nature. If time were actually isotropic, then there is a certain respect in which time order would not be an objective matter. For, if the laws of nature are time-symmetric, then for every law L formulated in terms of "earlier", there is a law $L^\star$, obtainable from L by replacing "earlier" with "later". But we may reduce our terms "earlier" and "later" to the expressions "earlier relative to $\overrightarrow{XY}$" and "earlier relative to $\overrightarrow{YX}$" where $\overrightarrow{XY}$ and $\overrightarrow{YX}$ refer to the future and past directions of time. And now we would be able to express the laws of nature without using a fundamental, *asymmetric* temporal relation. Rather, we could make do with relative priority—a relation that may itself be reduced to relations of temporal betweenness. The important point is that physics would employ no absolute temporal ordering relation, and this would constitute a sense in which time order is not objective. As William Newton-Smith (1980) puts it, there would be no under-lying asymmetric physical relation to which the observable quality of *beforeness* would reduce (e.g., in the way redness reduces to the emission of a certain wavelength). The explanation of our perception of beforeness would hinge on our temporal orientation with respect to the events in question, rather than any basic asymmetric relation between them. So beforeness would be, in a sense, not a wholly objective matter.

There is, therefore, *some* relationship between the anisotropy of time and the objectivity of time order. However, the concept of 'objectivity' is notoriously indefinite, meaning many different things to different philosophers. And so it is important not to confuse the relationship just described with another way in which objectivity and anisotropy have been linked together. I have in mind an idea due to Max Black (1959). The difference between his proposal and the one we have just examined is that Black ties the anisotropy of time to the thesis that our present, everyday conception of 'earlier' yields non-relative facts; whereas the previous proposal associated anisotropy with the ultimate need for a nonrelative time-ordering relation in formulating the basic principles of physics. Thus Black's thesis is:

> Time is anisotropic if and only if our ordinary statements of the form "x is earlier than y" are objectively true or false, and not merely true, or false, given an observer's point of view (neglecting relativistic considerations).

And since ordinary attributions of temporal order obviously are objective in Black's sense, he infers that time is anisotropic. What lies behind his criterion of anistropy is the idea that if time were perfectly symmetrical, then there could be 'time-reversed observers'—beings who are temporal mirror images of ourselves, who would have an opposite time sense to our own, and who would judge one event to occur earlier than another when we would place it later. Therefore statements of the form "x is earlier than y" would lack objectivity in just the same way as statements of the form "x is to the left of y" actually do.

However, this view of the matter cannot be right. Clearly *to the left of* is not an objective relation. Whether or not it holds is relative to a point of view. But nothing follows about the isotropic nature of space, or about whether the spatial mirror image of an arbitrary system would always be physically possible. This is an empirical matter which cannot be settled just by an examination of the logical properties of words. Indeed, there is experimental evidence against that empirical thesis. By the same token, even if *later than* were not an objective relation (even if its exemplification varied from one perspective to another), this would not entail that time is symmetrical. The most we might infer from such observer dependence is the nomological possibility of reversed observers with an opposite time sense to our own—'people' who become less grey and wrinkled as time goes on and who eventually turn into babies. And this possibility requires that for those processes, *ABCD*, involved in human development, the reverse sequences, *DCBA*, must also be possible. Such reversibility might suggest isotropy but would certainly not ensure it. For there could perfectly well be additional processes that are not reversible—phenomena having nothing to do with biology, and that play no role in human perception and growth. And their irreversibility would indicate time's anisotropy. Thus the nonobjectivity of *later than* would not entail the isotropy of time.

To see that the converse thesis also fails, suppose that there actually are some reversed observers, with an opposite direction of growth and an opposite time sense to our own, whose possible existence would be engendered by isotropy. These people say "*x* is later than *y*" in exactly the cases that we deny it. Now, it is not necessary to resolve the apparent conflict by treating such claims as nonobjective, on a par with "*x* is to the left of *y*". Rather, we might construe our *later than* assertions in such a way that a reference to *our* time sense is rigidly built into their meaning. For example,

> *x* is later than *y* *if and only if* the direction from *x* to *y* = the direction from *our* memories of events to OUR anticipations of those events

Similarly, the reversed observers might be construed in such a way that a reference to *their* time sense is rigidly built into the meaning of *their* expression. In that case they would not mean what we do by "later than". Each community would refer, using those words, to a different but quite objective ordering relation. The semantic properties of "later than" would resemble those of terms, such as "I", "here", and "now", and other words with an indexical aspect to their meaning.

The basic flaw in Black's account is fairly clear, I think. The question of time's anisotropy is, potentially, a subtle empirical matter—a question about the properties of an aspect of the world that is independent of human activity. The objectivity of our present notion of *earlier than* is, on the other hand, a matter concerning the meaning of an English expression. It was, to some extent, up to us to *decide* whether the proper use of "*A* is earlier than *B*" would allow us to accept both the affirmation and denial of that claim, when made from different perspectives. It is one thing to allow, as we saw at the beginning of this section, that "*A* is earlier than *B*" may be discovered to express a relation between *A*, *B*, and some implicitly specified temporal orientation; and it is quite another thing to say that it has no absolute truth value. Granted, if time is isoptropic, then its directions differ only with respect to their relations to other things. In that case our present claims of the form "*A* is earlier than *B*" would indeed reduce under scientific analysis to some statement relating *A* and *B* to a temporal orientation. In Newton-Smith's sense, time order would not be objective. But those claims, given what we *now* mean by them, would still be absolutely true or false. Admittedly, we might *come* to employ the term "earlier" for the relation that holds between *A* and *B* relative to a temporal orientation. But this would be to change what we presently mean by that word. Therefore, instead of inferring the anisotropy of time from the evident objectivity of our present concept of *earlier than*, Black should have been tipped off by his justified confidence in the latter fact, that this could not entail the anisotropy of time.

3. Grünbaum's reliance on de facto asymmetries

It would be a fantastic, though not impossible, coincidence for all the molecules of milk in a cup of coffee to have precisely the positions and velocities that would lead every one of them to move, within a short time, into the same small region. Such a thing *might* happen, but barely. The time reverse of milk spreading through coffee has never been observed, and never will be. Therefore, this process, though nomologically reversible, is said to be *be facto* irreversible.

Suppose that the only irreversible processes are, similarly, merely de facto irreversible, like circular waves expanding from a point, red-hot pokers cooling in the air, and milk spreading through coffee. The reverse sequences could occur, but just don't, because their initial conditions are so improbable. The question then arises whether we should conclude, from the prevalence of this sort of asymmetric process, that time is anisotropic? Do such phenomena mean that time itself is asymmetric?

Given the ideas developed so far, we must say that the answer to this question depends on whether the imputation of distinct causal roles to the directions of time is called for in order to explain such irreversibility. And the answer appears to be no. As we shall see in detail in the next chapter, there is no need to postulate any difference in nomological power between the relations *earlier* and *later* in order to explain why white coffee does not separate, waves do not converge, and so on.

A useful analogy may be drawn between time and space. Most people are right-handed and drive on the right-hand side of the road. These are de facto asymmetries in space. But as Earman (1967) has pointed out, no one is inclined to conclude, from those facts, that space is anisotropic. Similarly, we should not assume that every time-asymmetrical phenomenon is symptomatic of time's anisotropy.

In contrast with this point of view, Grünbaum (1963, 1973) does not accept that there must be time-asymmetric laws for time to be anisotropic. He says:

> The neutrality of our use of the term "irreversible" as between the nomological and the de facto senses is an asset in our concern with the anisotropy of time. For what is decisive for the obtaining of that anisotropy is not whether the non-existence of the temporal inverses of certain processes is merely de facto rather than nomological, instead what matters here is whether the temporal inverses of these processes always (or nearly always) fail to occur, whatever the reason for that failure! (1963, p. 211)

Thus, time certainly is anisotropic, according to Grünbaum, since there certainly are processes whose temporal inverses do not occur.

In support of this idea Grünbaum points out that it is hard to maintain a *sharp* distinction between laws of nature and other truths. We have definite intuitions that certain facts, such as the equal charge of all electrons, are laws of nature, and certain other facts, such as the radius of the earth, are certainly not laws. And there are borderline cases about which we are not sure what to say. For example, even if we knew that the universe had not existed for an infinite amount of time, it would not be clear whether or not to count this fact as one of the laws of nature.

But although the concept of law may be vague, this does not imply that the distinction between laws and nonlaws should be abandoned. Nor does it have the consequence that nomological and de facto irreversibility are on a par with respect to anisotropy. Rather, any vagueness in the notion of a law is transferred to the notion of anisotropy. If what is definitely a law is time-asymmetric, then time is definitely anisotropic. And if a fact whose nomological status is unclear is time-asymmetric, then it is similarly unclear whether or not time is isotropic. There is no reason, therefore, to suppose that any old temporally asymmetric feature of the universe—even one that results from a clear case of something that is not a law—should confer anisotropy upon time.

I suspect that the real source of Grünbaum's weak criterion of anisotropy is his rejection of *absolutism* with respect to time: he denies, like Leibniz does, that time is any sort of substantival entity over and above the phenomena that are located in time. This position leads him, mistakenly I think, to treat the question of whether time is isotropic as the question of whether the ordering of *states of the world* is isotropic. And, obviously, the answer to the second question is no. All it takes is the slightest de facto time asymmetry in the series of world states for the two directions along that series to be intrinsically different. Thus it is quite trivial that the universe is temporally asymmetric. Only the most bizarrely palindromic worlds would not be. However, there is no need to confound this fact with the anisotropy of time. Indeed, it would be wrong to do so, given what physicists and philosophers have generally had in mind. As we have seen, the question whether time is isotropic can plausibly be interpreted as a question about the intrinsic similarity between the past and the future directions. This may be construed as a question about the resemblance between the relations *earlier* and *later*—a question which does not presuppose a commitment to substantival time.

4. Can time change direction?

Is it possible that some parts of time are anisotropic and some parts are not? And could it be that the 'directions' of time in different regions conflict with one another? Our present analysis of anisotropy suggests that the answer to both questions must be yes. There could well be variable intrinsic differences between the two directions of time: intrinsic yet inconsistent differences that vary from one region to another with respect to both their existence and their sense. In order to clarify this general idea, I want to look at a specific theory of time that instantiates it, due to Reichenbach, and consider some criticism of this theory that has been offered by Earman (1974).

We shall analyze Reichenbach's view in much more detail in the next chapter. For now it is enough to note that he defines "the future" as the temporal direction of increase in the entropy of the majority of almost isolated ("branch") systems. I shall try to defend Reichenbach's idea against Earman's objection. But I should stress that my sympathies are with a single aspect of Reichenbach's approach, and not with the particular definition of "the future" that he advocates.

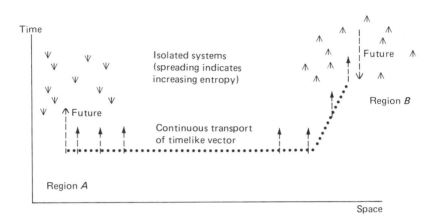

Figure 7

More specifically, I do not agree with him that our concept of the future reduces, either conceptually or empirically, to some entropic relation. Nor do I believe that the entropic asymmetry implies the anisotropy of time. But I think that Reichenbach is right in supposing that there *could* be an account of futurity that *does* imply a fluctuating anisotropy.

Earman points out that different regions of spacetime could well differ from one another with respect to the direction in which the majority of their systems display increasing entropy. Consequently, as Reichenbach recognized, different regions of spacetime may contain future directions that are, in some sense, 'inconsistent' with one another. Thus the direction said to be the future in region *A* may, when extended out of *A* and into region *B*, turn out (as depicted in figure 7) to conflict with the direction in which most of *B*'s systems display increasing entropy.

But this is unacceptable, according to Earman, for it conflicts with what he calls the Principle of Precedence (PP):

> Assuming that space-time is temporally orientable, continuous timelike transport takes precedence over any method (based on entropy, or the like) of fixing time direction; that is, if the time senses fixed by a given method in two regions of space-time (or whatever interpretation of "region" you like) disagree when compared by means of transport which is continuous and which keeps timelike vectors timelike, then if one sense is right the other sense is wrong. (1974, p. 22)

A *timelike* vector is one that points into either the absolute past or the absolute future. *Continuous* timelike transport cannot suddenly switch the direction in which such a vector is pointing. And temporal *orientability* entails, by definition, that as we transport along a closed curve in spacetime, beginning and ending at point X, continuity cannot determine that a future-pointing vector at X will "return" to X with the opposite orientation. Therefore, the restriction to temporally orientable spacetime guarantees that PP will not lead to inconsistency.

One bothersome feature of PP is the talk of "right" and "wrong" in reference to the determination of time sense. That gives the impression that if spacetime is temporally orientable, then there is a real problem of finding out which of the two possible orientations is *the actual one*. But this can concern only those who are in the grip of a 'moving *now*' conception of time. It has no bearing on the question of anisotropy: Is there a physically significant difference between the past and future orientations? Apart from metaphysical questions about the 'motion' of *now*, there are no further nontrivial issues as to which is *the* direction of time.

A second thing that should worry us about PP is that it seems insensitive to the possibility that time is merely locally anisotropic. Suppose that there is a physically significant difference between the past and future orientations. I don't see that we can rule out the possibility that this difference is not global but merely local to one or several distinct regions of spacetime. It is true that if spacetime is temporally orientable, then a particular orientation specified on one region will induce an orientation on any timelike vector. Thus, if an orientation is specified in some region on the basis of features peculiar to that region, this will nevertheless induce an orientation on curves in regions where those features are absent. But our concern is not simply to get a consistent orientation of timelike vectors. We can get that even if there are no physically significant differences between the two orientations. Rather we are interested in the nature of the temporal aspect of spacetime, both globally and locally. Given that its nature may vary from region to region with respect to anisotropy, and supposing that the details of the anisotropy might be (roughly speaking) of such a kind that it varies only with respect to existence and sense, why should we not mark these facts the way Reichenbach wants by relativizing the notions of "past" and "future" to regions of spacetime? (See Matthews 1979 and Sklar 1981.)

There is, it seems to me, a construal of Earman's PP that is not at all implausible; but Earman himself appears to reject it. Just as one of our criteria for being water might be, roughly, *having the same chemi-*

cal composition as this (pointing) *stuff*, similarly, one of our critieria for some temporal direction being the future direction at a given point could be *being the same direction as the direction in which these* (pointing) *branch systems increase their entropy*. Even if the term "future" is used in this way, it is still quite possible that certain regions of spacetime are temporally anisotropic, that some of these anisotropies have opposite temporal orientations, and that the anisotropy is intimately connected with the entropic behavior of branch systems.

Thus it may seem that Earman's PP can be reconciled with the major components of Reichenbach's view and will diverge only with respect to the semantics of "future". Whereas Reichenbach would let the direction to which it refers fluctuate with variation in the direction of an underlying anisotropy, PP would provide the notion with a sort of indexical element and require that the future direction be fixed by whatever is the sense of the anisotropy around here.

Earman's response is not accommodating. He says:

> But such an attempt doesn't even succeed in papering over the cracks. If the entropy method works for one region, why doesn't it work for all? If the resulting time direction depends on the choice of region to which the entropy method is applied, how do we know which is the right region to apply it to? If it is answered that there is no "right" region since there is nothing to get right or wrong, then entropy becomes entirely irrelevant—one might just as well flip a coin. (1974, p. 23)

It seems to me that what is betrayed again in these remarks is a unwillingness to accept that one might wish to use the words "past" and "future" to do something other than simply pick out the two distinct globally consistent time directions. Certainly one might decide to use those words for that purpose, and if so, Earman's PP will be satisfied. However, it is perfectly conceivable, and by no means obviously unreasonable, that we would wish the term "future" to designate the direction of an intrinsic asymmetry within spacetime— one whose orientation may vary from region to region. It seems to me that both decisions would be compatible with our present usage, and there is no fact of the matter as to which decision would be right. Moreover, I cannot see very much at stake here. If we do abide by PP and let the future direction at any point of spacetime be determined by which is the future direction here, then we are still quite at liberty to introduce new time-direction words in order to designate varying intrinsic temporal anisotropy. On the other hand, if we conform to Reichenbach's proposal, we nevertheless possess a pair of expressions, "our future" and "our past", that do conform to PP. And just as we often omit specification of a frame of reference when describing something's speed and are understood to be giving it relative to the Earth's surface, we might use the words "past" and "future" in accordance with PP and still reserve the right to wonder "whether or not the future relative to region B is the future."

5. Irreversible processes

Nearly everyone would agree (pace Grünbaum and Earman) that the anisotropy of time has a great deal to do with the existence of nomologically irreversible processes. But not enough attention is generally given to exposing the precise nature of the relationship between those ideas. In particular, it is no trivial matter even to say exactly what irreversibility would be.

A type of process, P, is nomologically irreversible if and only if the temporal inverse of P, $R(P)$, is incompatible with the laws of nature. That much is clear. However, the question arises (Earman 1967) as to what process would constitute the temporal inverse, or temporal mirror image, of an arbitrarily given process. We should provide some characterization of the function R.

A natural first thought will be that if process P is made up of the sequence of states, $ABCD$, then $R(P)$ is the sequence, $DCBA$. In general, one is tempted to suppose that $R(P)$ contains just the same events and states as P, but occurring in the opposite temporal order. However, this characterization must be rejected, for, on reflection, it clearly fails to capture what we have in mind by the inverse of a process. To illustrate, let P be the sequence, A (meteorite comes flying toward the Earth), B (hits the ground), C (bounces around), and D (stops). Surely, we don't suppose that the inverse of this type of process is $DCBA$—one in which a meteorite first stops, then bounces around, then hits the ground, and, finally, comes flying toward the Earth. Moreover, our problem is not resolved by restricting the constituents, A, B, and so forth, to instantaneous states. At any time the state of a moving object includes its velocity and position. So it is plain that the inverse process does not contain precisely that state, but rather a very similar state involving the same position with an *opposite* velocity (the same speed but in the opposite direction). The moral here is that if state A occurs in process P, then $R(P)$ contains, not A itself, but rather $R(A)$, the temporal inverse of A. It is plausible to suppose that when A is a state involving a specific velocity, the temporal inverse of A will involve the opposite velocity. However, we yet have no *general* account of how to construct the temporal inverse of an arbitrarily given state. We know merely that

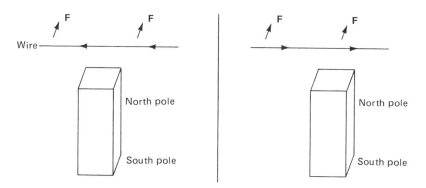

Figure 8

$$R\,(\ldots ABCD\ldots) = \ldots R(D)\ R(C)\ R(B)\ R(A)\ldots$$

which certainly *constrains* the notion of temporal inverse, but does not tell the whole story. Until this is rectified, we will not be in a position to determine, for an arbitrary process, what its temporal inverse would be; and this will hamper the investigation of the existence of nomological irreversibility and its relationship to the anisotropy of time.

The problem is not merely academic. For one can easily be mistaken about which process qualifies as the temporal inverse of a given process, and consequently misidentify cases of irreversibility and misjudge whether time is anisotropic. Sklar (1974) gives a good illustration of this. Imagine current flowing over a magnet, as shown on the left of figure 8. There will be a force **F** on the wire, at right angles to both the current direction and the magnet, pushing the wire into the page. We might well think that the temporal inverse of this process would be the one depicted on the right, and we might then conclude, from the fact that the right-hand arrangement is impossible, that the left-hand process is irreversible and that time is anisotropic. However, this apparent time asymmetry is just an illusion. When we understand how a magnet works—specifically that its magnetic properties are determined by tiny internal current loops of moving electrons—we can recognize that the right-hand diagram does not depict the true temporal inverse of the original process. Rather, in order to obtain a true inverse, the direction of the internal current loops would have to be reversed. Therefore, the magnet would have to be turned upside down, as shown in the modified right-hand diagram in figure 9. That process certainly is possible. And so we have here no threat to time reversal invariance and no indication of anisotropy.

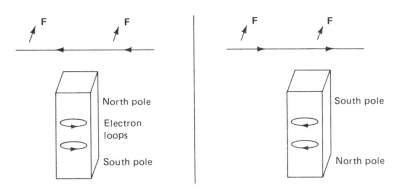

Figure 9

This example shows how the problem of identifying the inverse of a given process may be a subtle empirical matter. In order to solve it, it will be necessary to have discovered which mechanisms underly the process in question. Moreover we may extract from our convictions about such examples a crude, general characterization of what we have in mind by the temporal inverse of a process:

> Consider an arbitrary process, P, in a frame of reference whose spatial and temporal locations are represented by X and t, respectively. $R(P)$ is the process such that, for any fundamental particle or basic quantity e: e occupies or characterizes spatial region X at a time t in the course of $R(P)$ if and only if e occupies or characterizes X at time $-t$ in the course of P. Note that no quantity may qualify as "basic", in the required sense, if, like velocity, it consists in there being a relation between distinct states of the process at difference times.

This is illustrated in figure 10.

Once a genuine instance of nomological irreversibility has been identified, it is not hard to justify the inference that time is anisotropic. Suppose there is a physically possible process $ABCD$ whose temporal inverse is impossible. Let $(ABCD)$ designate the process whose temporal orientation is unspecified—merely that B occurs between A and C, and C between B and D. Then a physically necessary condition for the occurrence of $(ABCD)$ is that A is earlier than B. Thus the relation *earlier than* enters into explanations that are fundamental, for we have no deeper account of that necessary condition. In particular, we cannot suppose that the possibility of $(ABCD)$ will be found to depend on its orientation relative to certain other events: for in that case the reverse of $ABCD$ would not be physically impossible.

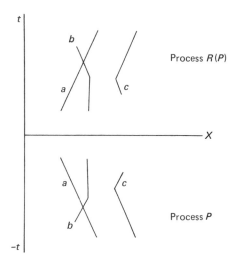

Figure 10

Therefore, it is plausible to suppose that there is some qualitative asymmetry in the temporal continuum that accounts for the fact that (*ABCD*) may be located in it one way round, but not the other—an asymmetry that will be manifested in some time-biased law of nature.

This line of thought suggests that nomological irreversibility is sufficient for anisotropy, but not that it is required. To resolve the question of necessity, we must ask ourselves whether the two directions of time could have distinct causal properties without this being reflected in the occurrence of irreversible processes. I think it is clear that the answer is yes. Time might be anisotropic in virtue of a time asymmetry in *statistical* laws. In that case it could be that the time reverse of every possible process is equaly possible, though not equally probable. Now this is not to accede to Grünbaum's view that mere de facto irreversibility is sufficient. For there to be anisotropy, the difference in probabilities between processes and their inverses must stem from the nature of time—which is indeed implied if the divergence is lawlike—and not be just a product of accidental boundary conditions.

6. The symmetry of time

Armed with this understanding of what it would be like for time to be anisotropic, we can finally consider the empirical question of whether it is so. The answer appears to be no. Despite a lot of looking, there have emerged no compelling reasons to adopt time-

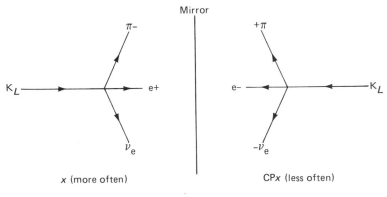

Figure 11

asymmetric laws (or to postulate de facto intrinsic differences between the past and future directions). Certainly, our fundamental physical theories do not incorporate a time bias. Even the notoriously irreversible phenomena of thermodynamics—processes of entropy increase which are typically associated with time's arrow—can, as we shall see in the next chapter, be reconciled with the isotropy of time. True, there is a difficulty with quantum mechanics. On some interpretations, *measurement* is treated as a basic irreversible phenomenon. However, the problems of how to make sense of the formal theory are themselves still so gigantic and intractable that no implications may yet be drawn from quantum mechanics regarding the anisotropy of time (Healey 1981).

The most substantial current challenge to isotropy comes, indirectly, from the observation of particle decay. Specifically, the neutral K meson (K_L) sometimes breaks up into an negative pion ($\pi-$), a positron (e+) and a neutrino (ν_e), as shown on the left of figure 11. But it also can disintegrate—though this happens less often—into a positive pion ($\pi+$), an electron (e), and an antineutrino ($-\nu_e$), as shown on the right. This alternate decay process appears to be the spatial mirror image of the first one, and constitutes its charge reverse. Therefore, the fact that the two processes do not occur with equal probability, though the difference is not large, indicates a failure of charge/parity (CP) reversal invariance. Now it is usually assumed, on plausible theoretical grounds, that *all* processes are charge/parity/time (CPT) reversal invariant. And this principle, together with the supposed violation of CP reversal invariance, entails a failure of time-reversal invariance. For if

Prob(x) $\neq$ Prob(CPx),   Experiment

and

$$\text{Prob}(x) = \text{Prob}(\text{TCP}x), \quad \text{CPT theorem}$$

then

$$\text{Prob}(\text{CP}x) \neq \text{Prob}(\text{TCP}x), \quad \text{Prediction}$$

We are led to expect, in other words, that the probability of a K_L having the decay products, $\pi+$, $e-$, $-\nu_e$, differs from the probability of a K_L being generated from those particles. And this suggests that CPx (the decay processes depicted on the right of figure 11) is not governed by time-symmetric laws.

However, this argument is far from airtight. First, the prediction has not been directly confirmed. And, even if it were true, it could turn out to be merely a *de facto* asymmetry, which does not involve time-asymmetrical laws of nature. Moreover neither the experimental nor the theoretical assumptions involved in the prediction are beyond question. For the frequency difference between the two forms of neutral K meson decay is not substantial and will perhaps be explained away. Anyway, the assumption that these processes are spatial mirror images may turn out to be false, as in the case of the magnets that we discussed earlier. Finally, the so-called "CPT theorem", though plausible, may be false. Since there are so many individually dubious assumptions in the argument, we may regard their conjunction as quite implausible.

Thus, not even this case—generally regarded as providing the strongest available evidence for anisotropy—gives us anything like a clear-cut failure of time-reversal invariance. Indeed, no such failure has ever come to light, despite our best efforts to find one. And this suggests that there simply aren't any.

Of course the fact that we have tried and failed to find something (e.g., a cure for cancer) does not always mean that there is no such thing. Our efforts, though falling short of complete success, may nevertheless indicate that success will be just a matter of time. Alternatively, we might have had some antecedent reason to think that the phenomenon would be exceptionally elusive or would be confined to areas that we cannot investigate. In that case our failure could have little epistemological import. However, the position with respect to anisotropy does not fall into these categories. There was no a priori reason to expect that science would not turn up a time-asymmetrical law, or that it would refute those apparent laws postulated to explain irreversible phenomena. There was no prior reason to think that any time-reversal-invariant laws would be confined to the domains of ultra-high energy or minute size that are difficult to explore. Nor has there been any discovery that raises the likelihood of irreversible phenomena in those domains, or that suggests any other intrinsic asymmetry between the past and the future. Thus our position is like that of the ornithologist who has looked thoroughly, but in vain, for a white raven and infers that there aren't any. Similarly, it seems fair to conclude, from what we know at the moment, that time is probably isotropic. Henry Mehlberg's (1962) old assessment still sums up the situation:

> There would be neither a miracle nor an unbelievable co-incidence in the concealment of time's arrow from us if there were nothing to conceal—that is, if time had no arrow. On presently available evidence time's arrow is therefore a gratuitous assumption.

4
Entropy

There are many natural phenomena we might call "one-way processes"—an evolving system undergoes a certain sequence of states as time goes on, but apparently never adopts that sequence of states in the reverse order. For example:

1. When hot and cold objects are brought together, their temperatures equalize to somewhere in between the initial temperatues. However, it virtually never happens, when two isolated objects are in contact at the same temperature, that one of them suddenly gets hotter at the expense of the other.

2. If a gas is concentrated in some small part of its container, it will expand so as to fill up the whole space available to it. But a gas that initially occupies the whole of its container does not spontaneously shrink into one corner.

3. A source of light will emit a spherical beam that radiates outward from the source. However, it never happens that a spherical beam converges inward toward a single point.

4. If a pack of cards is arranged in a very ordered state (e.g., all the black cards on top of all the red ones) and then thoroughly shuffled, the resulting arrangement of the cards will very likely have no recognizable order—the black and the red cards will be completely mixed up. But if, on the other hand, we start with such a mixed-up arrangement, it is extremely improbable that we would end with a highly ordered one.

The prevalence of such one-way processes might well suggest that time itself is asymmetric. For it is not unreasonable to suspect that it is the intrinsic difference between the past and future directions of time which is somehow responsible for the fact that there are such sequences of states which can occur in one order and not in the reverse order.

This idea is sustained within thermodynamics: a theory of heat developed in the nineteenth century by Carnot, Clausius, and Kelvin, to replace the caloric theory. The processes described in (1) and (2) fall within the province of thermodynamics. Their one-way nature is explained within this theory with the use of the notion of entropy—a measure, related to homogeneity, of how difficult it is to extract energy from a system—and by means of a law, the second law of thermodynamics, which is formulated in terms of this notion. The second law states that when any isolated system undergoes a process, its entropy can never decrease but must either remain the same or else become greater than it was initially. Given the association between entropy and homogeneity, this means that a system cannot move away from a state of uniform homogeneity. Thus, we can explain the irreversibility in (1) and (2). For the process of temperature equalization involves an increase in entropy. Therefore, since decreases in entropy are ruled out by the second law, the reverse of temperature equalization cannot occur. Similarly, the process of gaseous expansion involves a change toward an even distribution. Consequently the reverse process of contraction is forbidden, since it would involve an entropy decrease.

Thus we have a basis for time's asymmetry. For the second law says entropy cannot *decrease* but there is no law that says entropy cannot *increase*. Therefore the second law is not invariant under time reversal. And, as we have seen, this fact implies that time is anisotropic. For in giving different causal roles to the past and future directions of time, it implies an intrinsic difference between them.

2. Boltzmann's statistical mechanics

This situation was changed radically at the end of the nineteenth century when Boltzmann took vital steps toward reducing thermodynamics to statistical mechanics. His aim was to show how the thermodynamic behavior of a gas could be accounted for on the assumption that gases consist of billions of molecules that fly around in accordance with Newton's laws of motion. And he concluded that the second law is only approximately true—that the entropy of an isolated system *could* decrease (the gas *could* shrink into a part of its container) but that the probability of this sort of thing would be extremely low.

The theory of statistical mechanics has been continually reformulated, criticized and improved throughout the last hundred years. In what follows, I shall give an extremely simplified and informal rational reconstruction of some of those developments, focusing on foundational issues. The interested reader can consult Brush (1966) for translations of some important original papers and the Ehrenfests (1959), Klein (1973), and Kuhn (1978) for a fuller discussion of that work.

Consider an isolated system of particles—a gas—confined to a particular volume and possessing some total energy that remains constant through time. A *micro*state of the system is a fairly precise specification of the position and velocity of each particle. A *macro*state is a condition of the gas specified more crudely by thermodynamic parameters (e.g., volume and temperature) and other observable properties. Therefore each macroscopic state of the gas may be constituted by a certain number of alternative microscopic states. Now, Boltzmann assumed that every microstate is equiprobable. And this assumption has to some extent been supported by a series of mathematical papers—culminating in work by Birkhoff (1931) and von Neumann (1932) and contemporary results (reported in Arnold and Avez 1968)—that show on the basis of the laws of mechanics that (from *almost* all possible initial conditions and in certain very idealized circumstances) the gas will evolve in such a way that, in the long run, the relative frequency of each possible microstate is the same. These mathematical results are still very far from a proof that real systems in actual conditions will display an equiprobability of microstates. And certainly Boltzmann was not in a position to offer such a proof. For him, it was simply an empirical assumption (to be explained one day, perhaps, in terms of mechanics) that each microstate has the same relative frequency in the long run. Consequently the probability of any *macro*state would be proportional to the number of microstates consistent with it. Specifically, each macrostate would have a probability (relative frequency) equal to the number of microstates consistent with it divided by the total number of possible microstates. Boltzmann showed that the thermodynamic property of entropy was essentially nothing more than this probability, Thus the trend toward macroscopic states of higher entropy—that is, greater homogeneity—is explained as a tendency toward those macrostates that may be realized in relatively many microscopic ways.

To get an intuitive feel for this relationship between entropy (homogeneity) and probability (ease of microscopic realizability), imagine an enclosed volume of gas consisting of two connected chambers of equal size, as shown in figure 12. Let us compare the macroscopic states: "gas entirely in X" and "gas spread evenly between X and Y". Evidently, these two states differ considerably in entropy—one is much more uniform than the other. What we want to understand is how this difference corresponds to a difference in

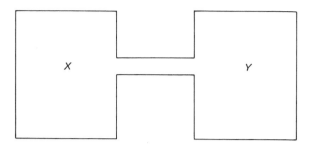

Figure 12

probability. That is to say, we want to see how the state of being confined to X can be realized in fewer microscopic ways than the state of being spread between X and Y.

One comparatively detailed method of describing the gas is to specify, for each particular molecule, exactly which chamber it is in. Of course there are alternative descriptions that are even more detailed. But let us regard the specification of a chamber, just for the sake of our example, as the characterization of a microscopic state. Each such state has the same low probability (equal to one half to the power of the number of molecules in the gas). So the probability of a macrostate is proportional to the number of microstates that realize it. Now, it is clear that many more such states will imply that the gas is spread between X and Y, than will imply that it is entirely in X. Indeed, there is only *one* such microscopic state in which every molecule is in X—namely, "molecule 1 in X, molecule 2 in X, molecule 3 in X, . . ." But there are an enormous number of specific arrangements of the molecules that can constitute the gas occupying both chambers. This is because, for any such arrangement, we can obtain an alternative just by interchanging one of the many molecules in X with one of the many molecules in Y—and each such swap engenders a new micro realization of the same macrostate, namely, the gas being spread between X and Y.

Boltzmann was able to show that, *in general*, macroscopic distributions of great inhomogeneity (high order, low entropy) could be constituted from relatively few microscopic arrangements of molecules, and were, therefore, relatively improbable. And he thought that an approximate form of the second law of thermodynamics would derive from these results. For, given the association of entropy and probability, that law would become the seemingly self-evident principle:

> Systems do not tend to go into states that are less probable than the states they are in.

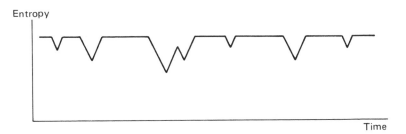

Figure 13

3. Reversibility objections

Despite the brilliance of Boltzmann's theory, his claim to have explained why the second law of thermodynamics is roughly true proved to be overstated. This was brought out, as follows, in some arguments inspired by Loschmidt (1876).

According to Boltzmann's statistical mechanics, all of the microstates of our gas are equiprobable—each possible specification of the position and velocity of every molecule has the same probability. But in that case each such microstate, B, is as likely as a corresponding state, B★, in which each molecule has the same position and the same speed as in B but is traveling in the opposite direction. Moreover, if microstate B is the potential *result* of an entropy-increasing process, then B★ would *initiate* the time reverse of that process, which would involve an entropy decrease. (I am presupposing that the microstates are so fine-grained that different systems in the same microstate will undergo a similar evolution, at least in the short run.) Thus any microstate—such as B—resulting from an entropy-increasing process should occur no less frequently than the time reverse of that microstate—that is, B★—a state which would produce an entropy-decreasing process. Therefore, contrary to our observation, and to the second law, entropy decreases should occur just as often as entropy increases.

Here is another way of putting the difficulty. The entropy of a system corresponds to the probability of its macroscopic state. And the probability of a state is the proportion of time (in the long run) that the system will spend there. It follows that the system will usually be in its high entropy (equilibrium) condition and will rarely fluctuate into low entropy states. Thus a graph of entropy against time for a permanently isolated system will look something like figure 13. Now we can see how this graph might be thought to imply a weakened form of the second law. For if we suppose that an arbitrary process is represented by an arbitrary portion of the graph, we can infer:

> An arbitrary process beginning at low entropy will most probably lead *to* a state of greater entropy.

But note that it would be misleading to represent this as a *slight*-modification of the second law; for whereas the second law is clearly not time reversal invariant, this new probabilistic version is. This version is quite consistent with the converse claim:

> An arbitrary process ending at low entropy was most probably derived *from* a state of greater entropy.

In fact, by referring to the graph, which is perfectly time-symmetric, we can see (1) how these converse claims are on a par with one another, (2) that we are unable to infer the temporal order of two states, given their respective entropies, and (3) that we as yet have no explanation for the phenomenon of one-way processes or for the approximate truth of the second law of thermodynamics.

4. Natural shuffling: Boltzmann's reply

Let us consider again the one-way process of gaseous expansion. A gas is concentrated in one corner of its container. Its entropy is a low $S1$. If left in isolation, the gas will expand until it occupies the whole container. In that state the value of the entropy will be a high $S2$. We can represent this process as

$S1$ . . . . $S2$

The double fact that we want to explain is that such processes often occur, whereas the converse processes

$S2$ . . . . $S1$

hardly ever occur. The reversibility objection to Boltzmann's statistical mechanics is not merely that the theory fails to explain this asymmetry, but that it incorrectly predicts that there should be no asymmetry at all.

This does not mean that statistical mechanics should be totally scrapped, however. Rather, the moral to be drawn is as follows. Clearly, Boltzmann's 'equiprobable microstate' assumption must, in light of the Loschmidt-style arguments, be given up in its *general* form. Nevertheless, the assumption is still plausible when *restricted* to the states of permanently isolated systems. And, indeed, such systems do not exhibit the one-way character that fuel the Loschmidt objections. Therefore, we may suppose that whatever the exact explanation of entropic phenomena might be, it is likely to have something to do with the fact that the systems undergoing one-way processes are *not* perpetual and *not* isolated systems. In other words, a natural response to the reversibility objections is a certain retrenchment. Let Boltzmann's equiprobability assumption apply only to certain idealized systems—permanently isolated ones—and let us explain the second law in terms of inevitable discrepancies from that ideal.

Departing for a moment from our chronological exposition, it is worth noting how this fact—specifically, the impossibility of complete isolation—is exploited in a modern approach to the problem, advocated by Lebowitz (1955), Blatt (1959), Morrison (1966) and Earman (1974). Their idea is that the entropy asymmetry exists because even those systems that we regard as "almost isolated" cannot escape the reach of gravitational fields and other interfering forces. Thus, let A be an initial fine-grained microstate in which a gas occupies only a small part of its container, and let B be the particular high entropy equilibrium microstate that ensues. By $A^\star$ and $B^\star$ understand the states that result when all the velocities in A and B, respectively, are reversed. In circumstances of complete isolation we could say that since A results in B, then $B^\star$ would result in $A^\star$. Moreover, if all microstates are equiprobable, we would expect the entropy-decreasing process, $B^\star \Rightarrow A^\star$, to occur as often as the entropy increasing $A \Rightarrow B$. And this is precisely Loschmidt's objection. However, since actual systems are, in fact, *not* isolated, but weakly coupled with external forces, there is no reason to expect that $B^\star$ would, in the real world, produce either $A^\star$ or any other low entropy state. This is because the precise organization of molecules required to evolve into a low entropy state is very sensitive, and would be wrecked by the slightest interference. Thus, outside disturbance destroys reversibility. Even if all high entropy microstates—including B and $B^\star$—occur equally often, we cannot conclude that for every final state of an entropy *increasing* process, there corresponds an initial state of an entropy *decreasing* process. Thus Loschmidt's objection appears to be dealt with.

However, it seems to me that although the impossibility of total isolation does deflect Loschmidt's argument, it does not completely block it and does not suffice to explain the entropy asymmetry. For, even in the presence of external influences, there could well be *some* high entropy microstate, C, on any given occasion, that would, given the specific external influence, I, present on that occasion, result in $A^\star$. In order words, granted that $B^\star + I \not\Rightarrow A^\star$. Nevertheless, how do we know that there isn't some other state C that compensates, so to speak, for the outside interference in just the right way to bring it about that $C + I \Longrightarrow A\star$? And if we continue to assume that all initial microstates are equally likely, then Prob C = Prob B. So it remains unexplained why entropy drops don't occur as often as entropy increases. Reference to external coupling has not provided the answer.

To put the point somewhat differently, the reversibility objections to the explanation of thermodynamics in terms of statistical mechanics arise from the thesis that all the microstates of a gas are equiprobable. Therefore, the problems cannot be solved without some way of seeing why it is that microstates are not in fact equiprobable when certain features of real systems are taken into account. But mere reference to the existence of weak interactions does not help. For it is far from obvious that the presence of slight external interference would favor some of a system's microstates over others. So we have been given no reason to expect that the graph of entropy against time for an *almost* isolated system would look very different from the time-symmetric graph for a *completely* isolated system, as shown earlier in figure 13.

What will help, however, is a recognition that there is a further respect in which the systems we actually observe are different from permanently isolated systems. This is their *short duration*. The systems we observe have not always existed and will not exist forever. They were brought into being; they interacted strongly with external influences at the time of their creation; and they will eventually be destroyed. Reichenbach coined the term "branch system" for these systems, which evolve in virtual isolation except for some initial or final strong interaction with their environment. A container of gas whose volume is suddenly increased is an example of such a branch system. While the gas expands to fill up the extra space, the system is more or less isolated. However, the first state in the process was the product of a strong external interaction. Some branch systems are connected with their environment only at one terminus. For example, in the preceding case the system might remain forever isolated after the initial interaction. Other branch systems may be connected at both termini. This would be the case in our example if, after some period of time, the container is destroyed.

Now, although in permanently isolated systems each microstate arises equally often in the long run (this is the equiprobability that we are now supposing is required by statistical mechanics), and although both drops and rises in entropy are equally likely developments in an eternal system (even if it isn't perfectly isolated), it does not follow (and moreover it is quite implausible) that in the *creation* of systems, the microstates that will lead to entropy drops, and those that will lead to rises, are brought about with equal frequency.

Therefore it is because of the prevalence of branch systems that one cannot be led, by statistical mechanics, to predict the nonexistence of one-way processes. We may respond to Loschmidt-style reversibility objections by pointing out that one should not identify the probability of a state in a short-lived branch system with the probability of that state's arising in a system that has always existed. Thus statistical mechanics gives no reason to expect that every *branch* system microstate will appear with the same frequency.

But all this is purely negative. We see that the reversibility objections may be dealt with by means of certain revisions in Boltzmann's initial conception, but it remains to provide a positive explanation of the second law of thermodynamics. Why is it that there are many branch systems whose initial entropies are lower than their final entropies, but there are almost no branch systems whose initial entropies are higher than their final entropies?

Boltzmann came to recognize that this question could not be answered by statistical mechanics alone. He then proposed that in order to explain the second law, statistical mechanics be supplemented with the assumption that our entire region of the universe has fluctuated into a state of relatively low entropy. The idea was that from this assumption it would follow that many branch systems with initial low entropy are created. Then statistical mechanics would take over to explain, first, why these systems become disordered, and second, why systems beginning with high entropy do not display drops in entropy.

Such an explanation of the second law is exactly parallel to the account one might give of the card-shuffling asymmetry mentioned at the outset. Why is it that there are relatively many cases in which shuffling a deck of cards, initially in a highly ordered sequence, produces a disordered sequence; whereas there are relatively few cases in which an initially disordered deck is shuffled into a highly ordered state? The answer is:

1. Shuffling is a random process which is equally likely to produce any of the possible sequences of cards.
2. There are many more disordered than ordered states.
3. Therefore (from 1 and 2) shuffling will very rarely produce an ordered state.
4. Prior to shuffling, a deck is often in a highly ordered state (as the result of deliberate arrangement).
5. Therefore (from 1 and 4) there are often cases of 'order → shuffle → disorder'.

6.   Therefore (from 3 and 5) there are many more cases of 'order
     → shuffle → disorder' than cases of 'disorder → shuffle →
     order'.

Boltzmann's strategy is in effect to treat thermodynamic processes
analogously:

1.   A gas evolves by a random fluctuation of molecules, which
     is equally likely to produce any microstate.
2.   There are many more microstates that engender macrostates
     with high entropy than low entropy.
3.   Therefore (from 1 and 2) gases will very rarely evolve into a
     state of low entropy.
4.   Initially, systems are often in low entropy states (because
     our region of the universe is currently in a state of low en-
     tropy).
5.   Therefore (from 1 and 4) there are often cases of increasing
     entropy branch systems.
6.   Therefore (from 3 and 5) there are many more cases of in-
     creasing entropy branch systems than cases of decreasing en-
     tropy branch systems.

5. Reichenbach's refinement

Reichenbach (1956) was one of the first to appreciate fully an important defect in Boltzmann's explanation. Statistical mechanics does not
imply premise 1, that a gas can be treated as if it evolves by random
fluctuation of molecules. Therefore statistical mechanics alone gives
us no right to expect that a system created in a state of high entropy
will not become ordered as time goes on. To see this, remember that
the theory of statistical mechanics is perfectly time symmetric.
(Otherwise, it could not possibly be reducible to Newtonian mechanics.) Consequently, if it precludes a certain process, then it must
preclude the time reverse of that process. But evidently, the theory
has nothing against the familiar fact that systems *terminating* in states
of high entropy often *begin* in states of low entropy. Therefore, as far
as statistical mechanics *alone* is concerned, it is equally permitted that
systems *beginning* in states of high entropy will often *terminate* in states
of low entropy. Therefore Boltzmann's assumptions must be further
supplemented in order to be able to justify premise 1 and thereby
account for the second law.

Reichenbach's positive suggestion for what the extra assumption
should be is not absolutely clear. His work on this subject was pub-

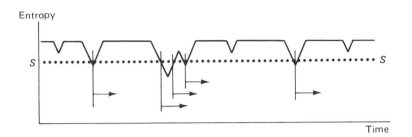

Figure 14

lished posthumously, presumably without the benefit of final revisions. My impression is that he wished to add the following crucial element to Boltzmann's view. In the initial attempt to account for the second law, it was assumed that the entropic behavior of an arbitrary branch system could be represented as an arbitrary segment of the entropic behavior of a permanently isolated system. But this idea was refuted by the second form of Loschmidt's reversibility objection. Therefore, suggests Reichenbach, we should retain the idea that branch systems may be regarded as arbitrary segments of perpetually isolated systems, but only in the following special, time-asymmetric sense: that the set of branch systems *beginning* in a macroscopic state of entropy S has the same entropic characteristics as the set of all the segments (illustrated in figure 14) that *begin* with entropy S in the evolution of a perpetually isolated system. The effect of this time-asymmetric assumption is to certify Boltzmann's premise 1. It allows us to suppose, in effect, that processes develop in the *future* direction by random fluctuation. It is this assumption that plugs the hole in Boltzmann's account, explaining why there are hardly any entropy-decreasing processes. For, whatever entropy we begin with, the vast majority of segments of the graph beginning at such a level will end at a similar or higher level.

Notice that the assumption is importantly time-asymmetric. We are given no analogous license to suppose that the states *preceding* a given state 'retro-develop' randomly. On the contrary, this is incompatible with Reichenbach's account. For he takes over from Boltzmann the idea that the universe has fluctuated into a low entropy state in which there are many low entropy branch systems. This idea, together with Reichenbach's assumption about evolution into the future, implies that the branch systems involving low entropy undergo entropy increases. But the time reverse of Reichenbach's assumption would rule out the occurrence of such processes, in just the way that his actual assumption rules out the existence of entropy-decreasing processes. Therefore Reichenbach's assumption precludes the adoption of its time reverse.

6. Grünbaum's formulation

Reichenbach's general approach to the explanation of one-way processes has been streamlined in the work of Grünbaum (1963, 1973). In the first place, Grünbaum is dubious of the idea, which Reichenbach borrows from Boltzmann, that the observable universe as a whole has some determinate entropy. He therefore does not embark on the attempt to explain why there are many branch systems with initially low entropy. He simply assumes that the universe is homogeneous with respect to this feature.

Second, he adopts a clearer version of Reichenbach's extra assumption. Grünbaum's formulation is that the initial microstates of branch systems are randomly distributed among the possible microstates compatible with the initial entropy. Now, if the initial state of a branch system has a high entropy, then from among all the possible microstates compatible with that entropy, only a negligible fraction could determine a significant reduction in entropy. Therefore the random distribution hypothesis implies that such entropy drops will almost never happen. For similar reasons, a system whose initial state has a low entropy (given Grünbaum's first assumption, there will be many such systems) will go through an entropy increase, leveling off at equilibrium in the vast majority of cases.

Grünbaum's theory is certainly quite plausible and attractive. However, there are a couple of respects in which it falls short of being a fully adequate account of one-way processes. In the first place, the fact concerning the commonness of low-starting-entropy branch systems, though evidently true, does not seem quite fundamental enough to constitute the basis of our explanation. We would like to know *why* it obtains. In the second place, the "random initial-microstate" hypothesis also requires explanation. Moreover, unlike the other assumption, it is far from obvious. So it would be desirable to find some independent reason for believing it.

One might have been tempted to look to quantum mechanics for the source of observed irreversibility. However, Grünbaum builds a convincing case—citing work by Watanabe (1935), Schrödinger (1950), and Rosenfeld (1955)—to show that an explanation of the second law should not be expected from the quantum theory. The main elements in his argument are (1) that quantum mechanical laws are, in the relevant sense, 'reversible' with respect to time, (2) that the one-way processes we see take place at a larger scale of magnitude than that at which quantum mechanical considerations come into play, and (3) that these processes exhibit their one-way character regardless of whether or not they are being observed.

7. 'Cosmic input noise' and 'initial micro chaos'

A view developed by Tolman (1934), Gold (1962), Davies (1974), Layzer (1975), and others to account for the frequency of low-entropy starting-states is to explain it in terms of the big-bang origin of the universe. The idea, very crudely, is that in the beginning the expansion of the universe was faster than the rate of equilibration, thus creating pockets of low entropy. In other words, the tendency of energy to even itself out was overwhelmed by the violence of the explosion, and so local concentrations of energy were formed.

This explanation of the first of Grünbaum's assumptions looks reasonably plausible, at least qualitatively. However, his second assumption is not so easily explained or validated. Other than its providing a neat account of the second law, there seems little reason to believe it; nor is there any obvious way of explaining it.

In the face of these difficulties perhaps we should try an alternative strategy. Let us dispense with Grünbaum's system by system 'random initial microstate' assumption and try instead to make do with a condition we might call "cosmic input noise", involving the global randomness of continuous, weak, external forces, to which every system is subject, and explained by a microscopic randomness in the initial conditions of the universe. Notice that this is a *time-asymmetric* condition: it is the input to a system, not its output, that is said to be random. The point of this idea is that even if, contrary to Grünbaum, the probability distribution among branch-system initial-microstates with a given created macroscopic character is far from uniform, this will not help the chances of a significant entropy drop. For the random nature of continual disturbing forces will overwhelm any systematic bias in the distribution of microstates produced by the "creating" initial interaction. And as long as this distribution is not correlated with the outside noise—as long as the internal state of the system and the external interference are not miraculously co-ordinated with one another—then the probability is very small that on any given occasion the initial state will be the one required to generate an entropy drop.

This position is closely related to the strategy of Lebowitz et al., discussed earlier, of blaming irreversibility on the fact that systems are never perfectly shielded, but always acted on, at least weakly, by external influences. I complained at the time that this account neglects the finite duration of branch systems and neglects the role of *creation* in explaining the prevalence of low entropy microstates. But there is a further reason why it is not enough to cite the everpresence of weak external coupling. A crucial element that this leaves out is the *randomness* of the disruptive influence. Without this condition, there is nothing to prevent the external forces and internal states from being coordinated with one another in such a way as to produce a low entropy state. To see what this would be like, simply consider the universe from the opposite temporal perspective to our own. What we normally see as the gradual decay of an ordered system will, from that point of view, look like the evolution of order from a miraculous combination of internal and external factors. There will seem to be plenty of order-'generating' processes—despite the constant presence of external 'inputs'. This can happen because the time-reversed disruptions, unlike normal cosmic input noise, are not random. So if the impossibility of complete isolation is to replace Grünbaum's assumption in the explanation of irreversibility, it is vital to stress the randomness of the external interactions.

The rough picture that emerges from our attempt to accommodate one-way processes is that of a universe that initially contains large-scale inhomogeneities of energy (macroscopic order) but that, subject to that constraint, is as chaotic as possible on the small scale (microscopic disorder). Then, as time goes on, the energy distribution becomes more uniform through processes of entropy increase (the macroscopic order decays in branch systems; the microscopic disorder produces 'cosmic input noise'). However, assuming that the basic laws are time-symmetric, then the products of these decay processes, though they seem chaotic on the surface, have a high *implicit* microscopic order, in the sense that they are the time reverse of states that cause macroscopic order. Thus we eventually obtain a final state in which all macroscopic order has been dissipated, and only implicit microscopic order remains. Unlike the initial state, this final condition is not microscopically random: from a time-reversed perspective it 'leads to' many branch processes in which macroscopic order evolves.

8. The fork asymmetry

In addition to its help with the second law, another attractive feature of the initial micro-chaos condition is that it may be employed to explain a further pervasive asymmetry: *genuine coincidences rarely occur*. In other words, given a strong correlation between events A and B,

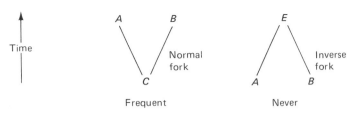

Figure 15

there is always some explanation—some earlier event, C—that causes them both. This fact is time-asymmetric, for it is frequently not the case that correlated events A and B have a characteristic joint effect E. Consider, for example, the expansion of a spherical light wave (or a circular water wave) from a source. In such cases A and B are distinct, yet perfectly analogous, wavefront segments, and C is the central disturbance that explains them. As Popper (1956) emphasized, the time reverses of such things do not happen. We never see inwardly moving spherical wavefronts converging to a point—an uncaused correlation. The general structure of this asymmetry—first characterized by Reichenbach (1950), and elucidated by Wesley Salmon (1984)—is represented in figure 15. Highly correlated events types are invariably preceded by some unified common cause, but they need not have a joint effect. We never find a pattern, as depicted on the right of figure 15, in which correlated events are linked only by having a joint effect.

Let me state this more carefully. Suppose that A and B are types of event (e.g., thunder and lightning) that are not often exemplified:

ASSUMPTION 1

$P(A)$ and $P(B)$ are small.

Suppose also that, given the presence of an event of one type, the probability is greatly enhanced that an event of the other type is also present:

ASSUMPTION 2

$P(A/B) \gg P(A) \cdot P(B)$.

In this case there will be a range of alternative event types, $C1$, $C2, \ldots, Cn$, whose instances tend to cause instances of A and of B. Note that each C-event is a 'unified cause' of A and B, in the sense that it cannot be split into two parts such that one causes A and the other B. Note also that A and B may sometimes co-occur without having any common cause—but this is rare. In contrast, there need be no range of event types, $E1$, $E2$, . . . , En, whose instances are almost always joint effects of A and B. Rather, it may well be that in a substantial proportion of cases in which A and B co-occur they have no joint effect.

This so-called "fork" asymmetry can be related to our discussion of entropy in a couple of ways. First, the condition of initial micro chaos, which helps to explain the second law, also explains the fork asymmetry. For an uncaused correlation of A and B could occur only if their causal antecedents were correlated; and this would eventually entail a correlation among initial conditions, which is inconsistent with the hypothesis of initial microscopic chaos. Thus, the truth of that hypothesis explains why the pattern depicted on the right of figure 15 does not occur. In other words, relative to the presence of a C-event, the chances of A and of B are much higher than they would otherwise be:

ASSUMPTION 3

$$P(A/C) >> P(A), \quad P(B/C) >> P(B)$$

But—because of the initial chaos condition—relative to the absence of any such C-event, A and B are statistically independent of one another:

ASSUMPTION 4

$$P(A\&B/-C) = P(A/-C) \cdot P(B/-C).$$

From assumptions 1 through 4 it follows that $P(C/A\&B) \simeq 1$. That is, nearly all co-occurrences of A and B are explained by the presence of a common cause C. Notice, by the way, what blocks the time reverse of this argument. We can legitimately substitute E for C in the first three assumptions, but if the fourth were changed in this way, it would be incorrect. Since the time reverse of the condition of initial chaos is false, we cannot suppose that the alternative to a joint effect, E, is some condition relative to which A and B would be independent of one another.

A further point worth mentioning is that thermodynamic irreversibility can be seen as a special case of the fork asymmetry. For we saw earlier that the nonoccurrence of entropy-decreasing branch systems is due to the absence of any correlation between the created initial conditions of the system and the forces impinging on it from outside. These two factors are not causally connected. And so their lack of correlation is simply an instance of the general phenomenon just depicted.

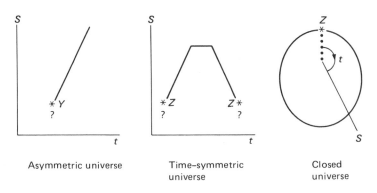

Figure 16

Supposing that the initial micro-chaos condition obtains, it would be desirable to find an explanation of it. Davies (1974) has suggested that the randomness could derive from the fact the universe originated from a singularity in spacetime. However, it is unclear how a transformation from spacetime singularity to random initial conditions could come about. Moreover there is a difficulty in reconciling this proposal with the possibility that the universe will come to an end, after periods of expansion and contraction, in a final singularity. If the universe will end in a singularity, it is hard to see how its beginning in one could account for temporally asymmetric phenomena. However, there is intriguing work by Schmidt (1966) and Cocke (1967) that hints at a possible solution to this problem. They have concocted cosmological models in which there is exactly the combination of randomness and implicit order in the initial conditions to bring it about that during the contraction phase (but not before) there is a profusion of entropy *decreasing* branch systems—a phenomenon that would not be guaranteed merely by the fact of contraction. These models (represented by the middle diagram in figure 16) have the satisfying property that, not only are the laws invariant under time reversal, but also the initial and final conditions of the universe are similar to one another. Thus the universe becomes much more thoroughly time-symmetric. However, it remains somewhat mysterious why such extraordinary boundary conditions should obtain. For they would comprise a miniscule proportion of the possible initial conditions. Reference to spacetime singularities isn't of much help, unless it can be shown how they would lead to the right sort of boundary conditions. A final possibility, worth contemplating, is that the mystery can be finessed if we simply *identify* what we previously called the initial and final states, and suppose that time itself is closed. This manoeuvre would have the advantage of leaving us with no irksome boundary conditions to account for.

In subsequent chapters I will occasionally invoke the initial micro-chaos hypothesis, and I shall assume that the final conditions of the universe are not random in this way but contain implicit order. Thus I shall write as though the asymmetric model is correct. But this is for the sake of convenience. To reconcile what I shall be saying with the other models, the term "initial chaos" could be taken to refer to that aspect of initial microstructure whose consequences are manifest during the expansion period, and the term "final conditions" could be understood as referring to states that just precede the period of contraction.

Finally, I should make it clear that nothing in what follows will depend on the speculative physical hypotheses that I have suggested here. What *will* be important for explanatory purposes is the fork asymmetry. And if my crude explanation of that phenomenon is along the right lines, so much the better. For we will then have an even deeper understanding of those things, such as our special knowledge of the past, that I shall be trying to assimilate to the fork asymmetry.

5
Knowledge

1. Biased information flow

It is evident that we know—or at least take ourselves to know—a great deal more about the past than the future. What is far less obvious, however, is why this is so, and even precisely what the phenomenon is that we somewhat crudely describe as "our knowing more about the past than the future." It is these two questions, concerning the nature and explanation of epistemological time asymmetry, that I wish to address in this chapter.

In forming judgments about the past, we have our own memories to go on, the testimony of others, records, history books, and many other types of trace of what has happened. On the other hand, our capacity to predict the future with any real confidence seems limited to very few contexts: notably, our own intended actions and the behavior of simple, isolated (or almost isolated) physical systems—behavior inferred from their present condition via more or less deterministic laws. As a consequence of these temporally selective avenues of information we know the names of the last ten presidents of the United States but not the names of the next ten, nor even that there will be ten more presidents. We know which cities we visited last year, but not, with the same certainty, where we will go next year. We can tell by inspecting the ground outside that there has been a rain shower, but it's hard to predict with equal confidence that there will be one. We can even claim to know that one hundred thousand years ago the Earth was inhabited by creatures of a specific kind, but we have no idea what, if anything, will inhabit the Earth in another hundred thousand years.

Struck by countless facts of this sort, one is inclined to conclude, in general, that we know more about the past than the future, and to begin casting around for an explanation. We must, however, bear in mind that this characterization—namely, that we know more about the past than the future—is far from satisfactory and may serve only as a crude preliminary indication of what is illustrated by our list of examples. What is it to know *more* about the past? Does this mean that the number of past facts known by each of us exceeds the number of future facts known? Surely not, for it is quite compatible with the phenomenon we have in mind that we know an infinity of future facts. How then are we to compare the quantities of past and future knowledge? Until this question is answered, our contention that we know more about the past than the future will remain somewhat obscure.

Nevertheless, I think our best policy is to proceed by searching for an explanation of the epistemological asymmetry and not to insist that this search be postponed until a more precise characterization is formulated. This is reasonable, first, because we have said enough about the phenomenon to provide a grasp upon what it is that we wish to explain. And second, we can expect that the proper characterization can only emerge in the light of a deeper understanding of how it comes about. This is not an unfamiliar situation in science.

I shall eventually propose an explanation of the knowledge asymmetry in terms of the fork asymmetry (the tendency for regularly associated events to have a characteristic antecedent but no characteristic effect), and therefore ultimately in terms of those cosmological facts that were invoked in the preceding chapter, in order to explain the fork asymmetry. However, before expounding this view, I would like to discuss briefly various competing suggestions.

2. Six purported explanations

Freedom of the will
One might suppose that the future is relatively unpredictable because of the combination of two facts: the past is fixed, in that nothing anybody does from now on can affect it; whereas the future is un-determined, depending in many respects on the exercise of free choice. There is no doubt some merit in this answer. However, it leaves out a great deal and cannot be regarded as the general explanation we are looking for. Not only should more be said about the relationship between freedom and unpredictability, but even given such an account, it seems clear that the answer will be too narrow. For there are lots of future events that are outside our control and nevertheless cannot be predicted. The weather is a notorious example. Perhaps human action is peculiarly unpredictable, and perhaps this contributes toward the difficulty of predicting many other future events. How-ever, it clearly is not the only factor, nor, as we shall see, is it even the predominant one.

Falsifiability asymmetry

A not wholly unnaturally thought about the matter is that we are not, and will never be, in a position to verify directly our claims to knowledge about the past; therefore we can afford to be overgenerous in our attribution of such knowledge, without fear of embarassment. On the other hand, we must be more careful in presuming to know the future, because we will be able to compare any prediction with the future facts. Thus our standards of justification for predictions must be, and are, more rigorous than our standards for retrodiction. That is why we take ourselves to know more about the past.

Alas, this line of thought is not only quite implausible—denying that we really do know more about the past—but it is also patently circular—explaining our belief that we know more about the past than the future in terms of that very assumption. To see this, note that it is taken for granted that predictions can be tested whereas retrodictions cannot. In other words, if I now make a claim about the occurrence of some event next week, then, when next week comes along, my prediction will be recollected and compared with the facts. On the other hand, if I make a claim now about the occurrence of some event last week, no analogous test is possible. For last week, when I would have been in a position to compare that claim with the facts, I had no idea that such a claim was going to be made. Thus our explanation is founded on the supposition that it is much easier to find out what has been claimed than what will be claimed. And this supposition is an instance of the very asymmetry that we are trying to understand.

Direction of causation

Earman (1974) has suggested that the source of epistemological time asymmetry is causal time asymmetry. Specifically, we know more about the past than the future because we have traces of the past and no traces of the future. And this is so because the present state of the world is caused by past events but not by any future event.

Now it seems to me that the first component of this explanation—we know more about the past because there are traces of the past—does no real work. We surely mean nothing more by "trace" than "event providing information about its causal antecedents." Thus the first component really amounts to the claim: we know more about the past than the future because the present provides information about its past causes but not about any future causes. Now we add to this what purports to be its explanation—namely, the direction of causation—but all that is really accounted for is why, given our definition of "trace", there are no traces of the future: the present provides no information about "its future causes", because there aren't any. In order to appreciate the gap in this account of the knowledge asymmetry, remember that a phenomenon may provide information about its effects, even though it cannot be described as a trace of those effects. And nothing has been said about why this type of epistemological access is less common or less reliable than epistemological access by means of traces. In other words, it remains to be shown why the present provides more information about its causes than about its effects. This would need to be done if the explanation in terms of causation is to be adequate.

This is not to deny that there may well be an intimate association between the causal asymmetry and the knowledge asymmetry—an association with the consequence that if there were a great deal of backward causation, then the knowledge asymmetry would be diminished. My point, rather, is that it is incumbent upon someone who proposes such an account to articulate the association and to explain why it is so much easier for us to infer the causes of an event than its effects.

Causal theory of knowledge
An alternative account of the asymmetry in terms of causation might be thought to derive from the so-called "causal theory of knowledge." That theory was proposed in response to various counterexamples to the classical analysis of knowledge as justified true belief. Gettier (1963) pointed out, for example, that someone may be looking in the direction of a distant, sheep-shaped bush, behind which an actual sheep is hiding, and therefore have a justified true belief that there is a sheep in front of him, even though he doesn't see the real sheep, and so doesn't really know that there is a sheep there. It was then suggested, by Goldman (1967) and others, that the classical conditions for knowledge be supplemented with a causal condition. Suppose, to take the crudest possible version, that we introduce a requirement that the fact known *cause* the belief in that fact. If this idea were correct, it would follow that only those facts capable of causing our beliefs in them may be known. And this, together with the future direction of causation, would entail a radical epistemological time asymmetry.

However, this particular strategy is defective in two respects. In the first place, the fact that it engenders such an extreme asymmetry suggests that there is something wrong with this crude version of the causal theory of knowledge. Surely we *do* know certain future facts that do not cause our beliefs about them. For example, we know that the sun will rise tomorrow. This means that though there might be a large grain of truth in the causal theory of knowledge, it is not quite correct as it stands. Indeed, its proponents usually advocate weaker formulations in which it is required merely that the fact believed be *causally connected* with the belief in that fact. But this improvement in the causal theory of knowledge destroys its time asymmetry, and thereby its capacity to explain why we know more about the past than the future.

Moreover, although we have tended to describe the epistemological asymmetry as a striking difference between our *knowledge* of the past and of the future, it seems plausible that the underlying asymmetry here has primarily to do with *justification* and with the reliability of mechanisms that link our beliefs to the world. Our disproportionate knowledge of the past comes from the fact that the procedures we use to arrive at beliefs about the past are generally more reliable than those generating predictions of the future. That phenomenon is not in the least bit illuminated by improvements in the semantic analysis of the word "knowledge".

Past-oriented overdetermination

Light emitted from a point is radiated in all directions, and the expanding spherical wave front may be divided into segments, each of which determines the time and place of the initial emission. Moreover, as we noted in the last chapter, this is a one-way process: its time reverse does not occur. David Lewis (1979b), generalizing from such phenomena, has maintained that all states of the world overdetermine their history, but not their future; and he argues (as we shall see in chapter 10) that this contingent fact will engender the future orientation of counterfactual dependence and thereby the predominant direction of causation. His overdetermination thesis, more precisely, is that for *every* event C there are, at every later time, events $E1$, $E2$, . . . , that each independently determine (given our laws of nature) that C occurred, but that the time reverse of this general fact does not obtain. That is, C may indeed be determined by earlier events, but not grossly overdetermined (there may be, at each earlier time, some event $B1$ that determines C's future occurrence, but no set of other events $B2$, $B3$, . . . , that also require C).

If this is correct (which is questionable), one may well hope to use it to explain the knowledge asymmetry (though Lewis does not do so). For it is plausible that the more symptoms of an event that are around, the easier it will be to detect one of these symptoms, so the easier it will be to come to know of the event.

However, as soon as this line of thought is spelled out, its defects become obvious. It is one thing to observe a symptom of some event;

but another, and much more difficult, thing to observe such a symptom and recognize it as such. The overdetermination asymmetry entails, at most, that the determinants of past events will probably impinge on our senses more frequently than the determinants of future events. But this will not account for the knowledge asymmetry unless we can take it for granted that those determinants are *known* to require the events they determine. However, not only is that presupposition quite implausible, it obscures the phenomenon that is really at the heart of the knowledge asymmetry. What makes it hard to know the future is not the scarcity of present phenomena that determine future events but rather the difficulty in discerning the future implications of what we do see. Moreover, the relative ease with which we can find out what has happened is not the product of the large *number* of traces of the past, but stems rather from their *quality*. In explaining the knowledge asymmetry, we need to reach an understanding of why there are no time-reversed versions of recording systems, such as memory, photographs, and books, that provide single reliable symptoms of the past. It is the absence of such high quality indicators of the future, not the lack of multiple determination, that is at the heart of the knowledge asymmetry.

Entropy
Reichenbach (1956), Grünbaum (1963), and Smart (1967) have elaborated the theory that our special knowledge of the past follows in a certain way from the (statistical) second law of thermodynamics. Their idea is this. There are many fairly isolated systems in highly ordered states. From our knowledge of our region of the universe, we can infer that such so-called "branch" systems did not achieve their degree of order by evolving in semi-isolation from less ordered states but, rather, evolved from some initial condition of similar or even higher order. That initial condition would have to have been produced by some external interaction with the system. The slowly evolving state of such a branch system provides us with a trace of this initial interaction.

In other words, the explanation of the epistemological asymmetry is that in virtue of our knowledge of the enormous preponderance of entropy-increasing over entropy-decreasing quasi-isolated systems in our part of the universe, we are able to infer when we see such a system in a state of low entropy (e.g., a footprint in the sand) that this state was preceded, and caused, by an interaction with the environment (e.g., a stroller). Thus the fact that we know more about the past than the future is held to be intimately associated with our awareness of de facto irreversible processes and with our belief in the approximate truth of the second law of thermodynamics.

I suspect that this point of view is widely accepted. However, it has been subjected to telling criticism, particularly by Earman (1974).

In the first place, recording systems, such as books, tapes, photographs, and brains, are generally in continual interaction with the rest of the world. Therefore it seems inappropriate to explain their behavior on the assumption that they are 'almost isolated'.

Second, in describing an informative state of a recording system as 'a low entropy state', this thermodynamic concept is applied beyond its unquestionably legitimate range. It was introduced for a specific theoretical purpose in order to help to explain the behavior of gases. Consequently there is substantial risk of incoherence in blindly applying the concept to quite different systems—like arrangements of playing cards, grains of sand on the beach, and marks on a piece of paper. Moreover, even if we could explicitly extend the concept of entropy in a natural way, so that it can be employed in the characterization of nonthermodynamic systems, we would still have no right to assume (1) that some principle like the approximate second law of thermodynamics would govern the behavior of 'entropy', in the new sense of the word, and (2) that even if the second law did apply, it would do so for the same underlying reasons as the original second law.

Third, the capacity to infer merely that the present state of a system was caused by some earlier interaction doesn't do justice to our ability to find out in detail what has happened in the past. For example, a photograph tells us a lot more than the mere fact that the paper has undergone an external interaction. But such extra information is left wholly unexplained by the approach under consideration.

And fourth, the inferences we actually make from present events to past circumstances often seem to have almost nothing to do with considerations of branch systems, order and entropy. When a bomb is dropped on a city, the explosion may leave traces. From the ruined buildings we may infer that a bomb went off. But in what sense would this be formation of a branch system in a state of high order and low entropy? On seeing footprints in the sand, we may infer that someone was walking there. However, if we see no footprints, we may similarly infer that no one was walking there. So even if we grant the unclear assumption that footprints constitute a highly ordered state, it seems as though our ability to make inferences about the past does not depend on the observation of such highly ordered states.

3. Against futurology

Having considered and rejected six different explanations of the knowledge asymmetry, it is now time for something more constructive. I think we should start by looking at the question in a relatively less abstract way, and concentrating our attention on some particular concrete phenomena that give us knowledge of the past. A reasonable list would include memory, writing, photographs, tape recordings, footprints, fossils, and paintings. The problem then is to describe the general type to which such recording systems belong and to explain why the time reverse of that type of system is not exemplified.

I shall divide this project into four stages: first, to give a highly idealized picture of a typical recording system; second, to indicate some respects in which actual systems depart from this ideal; third, to identify general facts that explain the presence of such systems; and, finally, to show why the time reverse of recording systems do not exist, by reference to the irreversibility of the general facts whose presence will have been found to be necessary for the existence of ordinary recording systems.

To begin, let me try to give a characterization of an idealized recording system by generalizing from the list of examples and neglecting the features of those systems that detract from their performance. An ideal recording device is a system, S, with the following characteristics:

1. S is capable of being in any of a range of mutually exclusive states $S0$, $S1$, $S2$, ...
2. Except for $S0$, these states are perfectly stable; that is, if S is in state Sk at time t, then S is in state Sk at all times later than t.
3. There exists a range of mutually exclusive external conditions $C1$, $C2$, . . . , to which S is sensitive in the following sense: if S is in its 'neutral' state $S0$ at time t, and the external condition Ck obtains in the environment of S, then S will go immediately into state Sk; moreover this is the only way that Sk can be produced.

Thus, if a system S is observed to be in state Sk, it can be inferred that at some earlier time the conditions surrounding the system were Ck.

Second, we should note certain respects in which actual recording systems fail to conform fully to this ideal. It is particularly important, as we shall see, that real systems do not perfectly satisfy conditions 2 and 3. There is generally a fair degree of stability. Nevertheless, it can happen, and often does, that once in a state other than $S0$, a recording system will be knocked out of it into a different state by some further external interaction. Moreover, the new state may be brought about by factors other than its 'canonical' cause. In other words, S may be knocked into Sk by factors other than Ck. A significant implication of these departures from the ideal is that we cannot now be absolutely sure, upon observing the system in state Sk, of the earlier occurrence of Ck. For it may have been that Cj occurred in the vicinity of S, putting it into Sj, but then some subsequent factors may have shifted the system into Sk. Or, it may sometimes happen that the system is moved directly from $S0$ into Sk by external factors quite distinct from any of the special conditions $C1$, $C2$, ... Therefore the observation of Sk does not conclusively establish the prior occurrence of Ck. Nevertheless, and this is a crucial point, it does make Ck very probable. For we eventually come to notice that the vast majority of cases of Sk were preceded by Ck; and so we infer the prior presence of such circumstances whenever Sk is observed.

In the third place, having described the way in which non-ideal recording systems are used to infer past facts from present traces, we can try to explain *why* there exist systems with the requisite properties—that is, we can try to uncover those general features of the world that are necessary if we are to acquire knowledge of the past. To this end, it is important to recognize that the phenomenon of recording is an instance of the pattern of events that is known (see chapter 4) as a "normal fork". Let me first sketch this idea, and then go over it in more detail. A recording system, S, gets into each of its informative states, $S1$, $S2$, . . . , much more often than it gets into its noninformative states—those that are not associated with any particular environmental circumstances. And this heavy clustering constitutes a correlation that is explained by the frequent presence of prior circumstances $C1$, $C2$, ... Thus, the association of S being in informative state Sk and prior condition Ck, which is essential to the performance of recording systems, is an instance of the general fact that correlations are causally explicable. Consequently, we would expect the explanation of that particular form of association to be the same as the explanation of the general causal connectedness of correlated events. And that, as we saw in chapter 4, hinges on the everpresence of 'cosmic input noise'—the randomness of background conditions. Let me amplify and clarify this proposal.

The generalizations that we rely on to gain knowledge of the past—namely, those of the form "Sk is usually preceded by Ck"— depend on the unlikelihood of anything other than Ck causing Sk. And this is explained, in turn, by the following two facts. First, the informative states of a recording system occur more frequently than its other states. And, second, apart from the tendency of certain conditions $C1$, $C2$, . . . , to cause certain states $S1$, $S2$, . . . , the effect of all other influences is random. In other words, background conditions I (other than $C1$, $C2$, . . .) are equally likely to produce any state of S; and Ck is just as likely to produce any of the states other than its canonical effect, Sk. These two facts imply that Sk is nearly always caused by Ck; for there is no other way for the preponderance of Sk to be accommodated. Here is a more formal version of the explanation.

The two basic facts are, first, the relatively high frequency of informative states

$$P(Sk) >> P(Sz) \tag{1}$$

where Sz is an arbitrary noninformative state, and P represents empirical (*non*subjective) probability and, second, the randomness assumption

$$P(Sz/-Ck) = P(Sk/-Ck) \tag{2}$$

In addition, we obtain from the probability calculus

$$P(Sz) = P(Sz/-Ck)P(-Ck) + P(Sz/Ck)P(Ck) \tag{3}$$

Substituting (2) into (3), we get

$$P(Sz) = P(Sk/-Ck)P(-Ck) + P(Sz/Ck)P(Ck) \tag{4}$$

Therefore

$$P(Sz) > P(Sk/-Ck)P(-Ck) \tag{5}$$

From (1) and (5), we obtain

$$P(Sk) >> P(Sk/-Ck)P(-Ck) \tag{6}$$

But from the probability calculus

$$P(Sk) = P(Sk/Ck)P(Ck) + P(Sk/-Ck)P(-Ck) \tag{7}$$

Therefore, from (6) and (7), we infer

$$P(Sk) \simeq P(Sk/Ck)PCk) \tag{8}$$

Therefore

$$\frac{P(Ck)P(Sk/Ck)}{P(Sk)} \simeq 1 \tag{9}$$

But Bayes's theorem entails

$$P(Ck/Sk) = \frac{P(Ck)P(Sk/Ck)}{P(Sk)} \tag{10}$$

Therefore, from (9) and (10), we infer

$$P(Ck/Sk) \simeq 1 \tag{11}$$

That is to say, Sk is a reliable symptom of Ck.

In short, random external interference is very unlikely to produce one of the informative states of a recording system; therefore any such observed state was probably produced by its normal cause. For future reference, let me emphasize a certain feature of the world that has turned out to be crucial in explaining the existence of recording systems. This is the fact that, apart from the circumstances, C1, C2, . . . , all other external influences on the recording system are *random* and have no tendency to produce any particular states. If it were not for this randomness of distorting influences, then there would be no reason why the systems proclivity for certain states S1, S2, . . . , would have to be explained by the prior existence of characteristic circumstances C1, C2, ...

In the fourth place, having characterized a certain type of non-ideal recording system and described the general facts necessary for its exemplification, we are now in a position to formulate our asymmetry problem as follows: Why is it that the temporal reverses of such systems do not exist? In other words, why are there no 'pre-recording' systems, $S\star$, satisfying the conditions:

1. $S\star$ is capable of being in any of a range of mutually exclusive states S0, S1, S2, ...

2. Except for S0, these states are fairly stable; that is, if $S\star$ is in a state Sk at time t, then $S\star$ is probably in state Sk at all times *earlier* than t.

3. There exists a range of mutually exclusive external conditions E1, E2, . . . , with which $S\star$ is associated in the following way: if $S\star$ is in its neutral state at time t, and the external condition Ek obtains in the environment of $S\star$, then, *beforehand*, $S\star$ was in state Sk; moreover this is what usually happens following Sk.

Such a 'non-ideal time-reversed recording system', $S\star$, would enable us to predict the future. An observation of $S\star$ in state Sk would indicate, with high probability, that the conditions Ek will obtain in the vicinity of $S\star$ at some time in the future. In other words, $S\star$ being in state Sk would be expected, with high probability, to cause Ek to occur. Thus our question of why we have better epistemological access to the past than to the future reduces to the question of why systems like $S\star$ are nonexistent.

The explanation that I want to propose is obtained by looking at

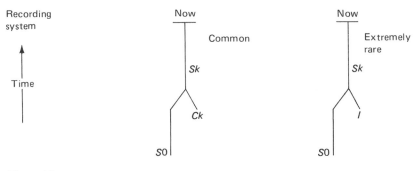

Figure 17

the general conditions that account for the existence of normal recording systems, and noting that the time reverse of those conditions do not hold. Now, I have suggested that the phenomenon of recording may be assimilated to the casual connectedness of correlated events. The rough idea was that if a system has numerous macroscopically similar states, of which some small number occur disproportionately often, then the tendency of the system to concentrate in those special states constitutes a correlation for which we should expect a causal explanation. That is to say, it is to be expected that there is some particular antecedent event that is correlated with the system being in one of the specially frequent states. However, there is no general condition that would similarly explain the existence of prerecording systems. There is no regularity to the effect that correlated events are always associated with some characteristic effect. And so there is no reason to expect that disproportionately frequent states of some purported prerecording system would tend to engender any particular types of result.

This explanation can be deepened by reference to the presence of random cosmic input noise, stemming as we saw in chapter 4, from the initially chaotic boundary condition of the universe. It is in terms of these phenomena that we were able to explain why, in the case of normal recording systems, the presence of state Sk was unlikely to have been produced by anything other than Ck (see figure 17). However, and this is the vital point, we cannot reverse this explanation because the time reverse of the 'initial micro-chaos' boundary condition does not obtain. Specifically, we *could* assume that an informative state Sk of a normal recording system would probably not arise from noncanonical factors, for those factors are random. But we *cannot* assume that such a state of a reversed system would probably not *decay* into a noncanonical effect, for there is no reason to expect

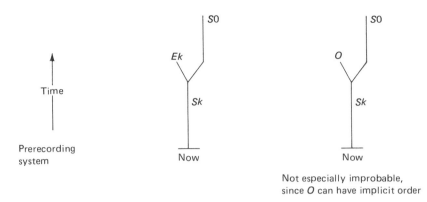

Figure 18

that the noncanonical effects of *Sk* will be random (see figure 18). Thus the environment does not spontaneously produce a misleading record (because of initial micro chaos); but a misleading prerecord could easily obtain, as an informative state simply decays leading to implicit output order. This can happen, since, unlike *initial* conditions, the *final* conditions of the universe are not random.

The idea is that random cosmic input noise is very unlikely to produce one of the informative states of an ordinary recording system; therefore any such observed state *Sk* was probably produced by its normal cause *Ck*. On the other hand, we cannot argue analogously that such a state of a time-reversed system would probably lead to a normal effect *Ek*. For such a state will very often decay or be destroyed, perhaps giving rise to hidden order among the effects of the decay. Such hidden output order is not precluded because there is no time reverse of the initial micro-chaos boundary condition—there is no final micro-chaos boundary condition. Thus, because of random input cosmic noise, and the absence of the time reverse of this condition, the departures from the ideal of a recording system are much less substantial in the case of an ordinary recording system than in the case of any purported prerecording system, where such departures would be so great as to be fatal.

In conclusion, let me mention some of the respects in which my explanation of the knowledge asymmetry both resembles and diverges from the entropic approach of Reichenbach, Grünbaum, and Smart. The main points of similarity are that a record is treated as an ordered state of a system whose order suggests characteristic causal antecedents; and that the existence of such systems (and the nonexistence of time-reversed versions of them) is alleged to be affiliated with the second law of thermodynamics. These similarities are important, and in light of them, one might well say that the present proposal falls within the general strategy suggested by Richenbach et al. But the differences are worth stressing. In the first place, it is not assumed here that recording systems are thermodynamic branch systems, and there is no suggestion that records are 'low entropy states' governed by the second law of thermodynamics. Nor do I suggest that the inferences we make from the observation of records involves any reference to the second law. Instead, I argue that records are 'forks'—instances of the principle that correlations are causally explicable. And so any disproportionately frequent macroscopic states of a system will have characteristic origins. We eventually learn to associate these states with their origins, and this association is the basis of our inference. Thus I distinguish between the facts that are employed in drawing conclusions about the past and the general physical principles that explain why such facts are available to us. This model allows us to see how our knowledge of the past can amount to more than the thesis that the recording system "has interacted with its environment", and it allows us to understand how historical claims can be based on phenomena that are evidently not low entropy states of thermodynamic branch systems. In addition, I have tried to give depth to the analysis by reference to 'cosmic input noise' and 'initial chaos'. My strategy here applies the general explanation of forks that was sketched in chapter 4 to the particular case of recording systems. Even if this extension of the analysis were to prove incorrect, the initial and main point would still stand: namely, that our special knowledge of the past derives from the fork asymmetry.

In later chapters I am going to suggest that the asymmetries of causation, explanation, decision, and value all depend (to some extent and in one way or another) on the asymmetry of knowledge. Before doing this, however, I shall address two preliminary questions. Is it possible for a cause to occur later than its effect? And is it possible for a human being to travel backward in time? These affiliated problems are fascinating in their own right. But in addition the treatment of them will help to prepare the way for the attempt, in chapters 8 and 9, to account for the *predominant* direction of causation and explanation.

6
Backward Causation

Opponents of backward causation typically rest their case on a couple of distinct considerations. Their first point is simply the claim that the normal time order of cause and effect has been 'built-in' to our concept of the causal relation—that a cause is, by definition, an *earlier* determining condition. I'll call this the conventional predetermination objection. To its credit, it provides us with an answer to the question, "Could a cause be earlier than its effect?" and also with a straightforward explanation of that answer. For the structure of the causal relation is invoked to explain *why* backward causation is impossible, not only to show *that* it is. However, the conventional predetermination theory is far from uncontroversial. This is because it conflicts with Quine's (1951) general arguments against the possibility of any 'truth by definition' or 'truth by convention' and, in particular, because it does not square with our ability to *imagine* evidence that would produce at least some temptation to postulate backward causation.

For example, in Michael Dummett's (1964) story of the dancing chief, we encounter a remote tribe whose young men are periodically sent on a lion hunt to test their courage. While they are away, the chief performs ceremonial dances intended to cause the hunters to act bravely. Now, as surprised as we are by the tribe's belief in the *long-distance* effect of these dances, we are especially astonished to see that the chief continues to dance even after the hunting is all over—while the young men are on their way home—at a time when, one would suppose, it is too late to influence their behavior. But he can justify this policy on the basis of past experience. In previous years, he says, an occasional failure to dance during the hunters' return journey has been associated with a much higher incidence of cowardice. And these results have increased his confidence that the dances indeed have a retroactive effect.

Schematically this is a case where a later event L, is highly correlated with an earlier event *E*—which strongly suggests some sort of causal connection between them. But the causal antecedents of *L* are thought to be sufficiently well understood to preclude all the familiar types of causal explanation—either that *E* causes *L* or that they have a common cause. And this appears to leave us with no alternative than the hypothesis that *L* causes *E*—the later event causes the earlier one.

Such hypothetical scenarios give rise to the suspicion that the conventional predetermination objection to backward causation must be mistaken. For they suggest that it is a subtle question, perhaps an empirical question, whether or not backward causation occurs. On the other hand, the theory that time order is an analytic constituent of causation seems to entail that there couldn't conceivably be such a thing as evidence, even hypothetical, suggesting backward causation, any more than there could be evidence tempting us to suspect the existence of a week in which Thursday came before Wednesday.

In light of this difficulty it is customary to deploy a second objection to backward causation. I'll call it the bilking argument. Versions may be found in the work of Tolman (1917), Flew (1954), Black (1956), Pears (1957), Dummett (1964), Earman (1972), Mellor (1981), and many others. The rough idea is that any backward causation hypothesis would be necessarily refuted by the following experiment: repeatedly wait to observe the presence (or absence) of the alleged effect *E*, and then try to prevent (or produce) the subsequent, alleged cause *L*. If this policy is carried out, then *E* will often occur in the absence of *L*, and *L* will frequently fail to bring about *E*. So the backward causation hypothesis is false. If, on the other hand, the attempt to carry out the policy fails, this indicates that an agent's ability to produce *L* depends on the prior presence of *E*, which in turn means that *E* is a necessary causal antecedent of *L*. Thus, whatever happens, the hypothesis will be falsified.

Back in the jungle: suppose we challenge the chief to adopt a bilking policy—that is, to find out straightaway how the young men have performed, and then to dance if, and only if, it is reported that they were *not* brave. If the chief accepts this proposal, then, it seems, he will be forced to abandon his belief in the efficacy of dancing. For either he will dance when the men have been cowardly, or he will find himself unable to dance. In the former case the correlation between dancing and bravery will be destroyed. In the latter case he will conclude that his dancing is not, as he had previously thought, an action wholly under his control. On the contrary, the bravery of the men turns out to be a necessary causal precondition for dancing. Thus, it is not that he dances to *make* the men brave. Rather, by seeing if he is able to dance or not, he can *find out* whether or not they have been brave.

My plan for this chapter is to do three things. First, I want to look carefully at the bilking argument. Does it really show that backward causation is, in some sense, impossible? Nonexistent? Or merely improbable? Or perhaps the argument is simply no good at all? I shall arrive at none of these conclusions. Rather, I believe that its virtue is to underline the fact that backward causation would be associated with the occurrence of inexplicable coincidences (correlated events that are not causally connected). As for the import of this association, I shall suggest that it varies from case to case. Sometimes we will have to conclude that the particular backward-causation hypothesis in question is unlikely to be true. But in other cases the right conclusion would be merely that if we do accept the hypothesis, then a degree of conceptual awkwardness will result. Thus the bilking argument is by no means a general refutation of backward causation.

Why is it, then, that backward causation is almost never seriously entertained? The reason, I think, lies in the fact that although there are certain kinds of evidential situation, like those in Dummett's example, that are especially conducive to the postulation of backward causation; nevertheless it so happens that such circumstances do not arise. I shall try to describe these circumstances in general terms, show how they suggest backward causation, and explain why they do not actually occur.

Finally, I shall reexamine the predetermination model of causation, and indicate a version of the doctrine that can be reconciled with the existence of prima facie evidence for backward causation. This idea will be elaborated in chapter 8 when we come to consider why the *predominant* direction of causation is from past to future.

2. Robbing effects of their causes

The first thing to notice about the bilking argument is that the usual formulation—the one I have just presented—is seriously defective, and that bilking considerations actually have much less force than is suggested by that formulation. We shall see, given a more careful statement of the bilking considerations, that they weigh against backward causation only by means of their support for a certain preliminary conclusion, whose bearing on the possibility of backward causation requires close scrutiny. More specifically, what the bilking argument shows is that backward causation would engender *inexplicable coincidences*. Now this preliminary conclusion may, perhaps, provide the basis for a good objection to backwards causation. But it needs to be shown how this is so, and that turns out to be a rather messy business.

Suppose the chief accepts our challenge and carries out the bilking policy. That is to say, he dances when he discovers the men were cowardly, and he doesn't dance after finding they were brave. In my initial statement of the bilking objection, it was suggested that this possible result of the experiment would *falsify* the chief's backward-causation hypothesis, since it would undermine the association he has noticed between dancing and bravery. But this claim is overstated. It fails to accommodate the impressive correlation that preceded our meddling. And it overlooks the possibility that the circumstances conducive to the chief's retroactive ability are quite delicate. We mustn't forget that nearly every scientific generalization has exceptions and is expected to hold only in 'appropriate' conditions. So there need be nothing unusual or objectionably *ad hoc* in supposing that the chief's dancing sometimes causes the hunters to have acted bravely and sometimes doesn't.

However, although we see that the backward causation hypothesis is not straightforwardly *falsified* by successful implementation of the bilking policy, this result does pose an awkward problem. Why do the chief's actions not continue to have their 'normal' consequences? Why is it that his dancing fails to work (or proves unnecessary) precisely on the occasions when the bilking policy is in effect? How, in other words, can we explain the coincidence of (1) the bilking policy and (2) a combination of the absence of circumstances that are appropriate for the dancing to cause bravery and the presence of phenomena that will bring about bravery when the chief doesn't dance. We are left, it seems, with an embarassing uncaused correlation—an inexplicable coincidence.

Similarly, if the chief does *not* follow our instructions, either because he won't or can't, this result will not falsify the backward-causation hypothesis. As before, it will merely raise questions of explanation. Here, it is worth distinguishing two possible scenarios. First, there are cases in which failure to bilk is explained by the fact that the earlier event, E, is a nomologically necessary condition for the later event, L. Perhaps, given the young men's cowardice, it is required by strange laws of nature that the chief will not dance. Second, there are cases in which the failure to conform to the bilking policy is explained by a tendency for extraneous factors to intervene. The chief's bilking plan may be continually thwarted by a succession of distracting circumstances—like wars, floods, and snakebites. Let me say a little more about these two scenarios.

Regarding the first type of case, we are allegedly supposed to conclude (according to my original statement of the bilking objection) that E (e.g., bravery) is a necessary causal antecedent of L (e.g., dancing), and so we allegedly must concede that the association between E and L is explained in terms of orthodox (future-oriented) causation. But this claim is simply mistaken. The observation that L is never present without having been preceded by E is explained by the fact that L is a sufficient condition for E. This fact is compatible with both the orthodox and unorthodox causal hypotheses and does not favor either one of them. Granted, we generally opt for an orthodox causal hypothesis when we can find one that seems adequate. And it might be argued, on this basis, that we should select the hypothesis that E (bravery) causes L (dancing) over its unorthodox rival. However, although this line of thought may be perfectly reasonable in itself, it is not consonant with the rationale for the bilking argument. The point of the bilking argument was supposed to be to prove that there could be no exceptions to the usual practice of orthodox causal interpretation. But according to the present strategy, the bilking argument presupposes that future causation is always preferred. And if we can assume this, then we have no need for the bilking argument.

Another reason that one might be tempted to say, given the imagined outcome of the experiment, that E is a necessary causal antecedent of L, is that the situation may look analogous to certain everyday situations in which talk of backward causation would be ludicrous. Suppose, for example, that L' is the act of switching on a light and E' is the immediately prior state of the bulb being in good working condition. Here, L' is a sufficient condition for E', but it would be silly to think in terms of backward causation. Rather, by trying to perform L', we can find out whether E' obtained. Similarly it may seem, given that the bravery of the young men is a necessary condition for the chief's dances, that it would be equally absurd to postulate backward causation. Rather, the chief can find out if the men have been brave by seeing whether or not he can dance. Thus the merit of the bilking argument, it might be said, is to show that alleged cases of backward causation may be assimilated to everyday situations in which backward causation is out of the question.

The flaw in this analysis is that there is a crucial disanalogy between alleged cases of backward causation and the light bulb example. Part of the reason that we have no temptation to think that turning on the light causes the bulb to have been sound is that there is no particular association between *trying* to turn on the light and the bulb having been in good working order. But our evidence in the backward-causation case was strikingly different. We found that the chief could almost always dance whenever he wanted to, and so there was a high correlation between his trying to dance and the young men's bravery. There is nothing analogous to this correlation in the light bulb case. Therefore it is wrong to charge that our inclination to adopt the backward-causation hypothesis simply confuses 'acting to *cause* an earlier event' and 'acting to *find out* if the earlier event has occurred'.

We are considering the first type of bilking failure—failures explained by the fact that the earlier event E, is a nomologically necessary condition for the later event L. And I have been arguing that this scenario would not falsify the backward-causation hypothesis; for it would not compel us to adopt an orthodox causal interpretation. However, I do not intend to suggest that the unorthodox interpretation would be unproblematic. On the contrary, the hypothesis that L causes E—that, for example, the chief's dances cause the hunters' bravery—engenders a puzzle about explanation that is similar to the one that we were left with in the case of successful bilking. The problem here is that events, such as acts of bravery, have characteristic causal *antecedents*. They are already determined by prior circumstances independently of whatever might happen later. Therefore, if we suppose that, in addition to these causal antecedents, such events have *subsequent* causes, then we are supposing that these events are regularly causally overdetermined; we are recognizing an association between the earlier determinants of E (e.g., childhood training) and the later determinants of E (dancing). But, according to the unorthodox interpretation, there is no causal connection between these correlated phenomena. So we are stuck, as before, with an unexplained coincidence.

The second type of scenario in which the bilking policy is not successfully carried out leaves us in somewhat the same boat. This is the sort of case where failure is due to a series of interfering circumstances. Such an outcome is very unlikely; for the association between intending to bilk and events (wars, illnesses, storms, etc.) that frustrate those intentions would be a pure coincidence. However, my point is that we have here again a possible result which, though problematic, does not straightforwardy falsify the backward-causation hypothesis.

Thus we see that the bilking objection is not, as originally advertised, that any backward-causation hypothesis could be automatically refuted by doing a certain experiment. The real point is this. If certain types of backward-causation hypothesis were known to be true, then one or another type of inexplicable coincidence would result from the bilking experiment.

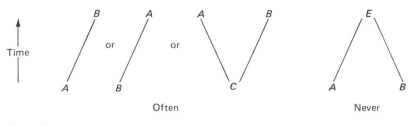

Figure 19

3. Inexplicable coincidences

The question remains as to what this conclusion implies for the acceptability of backward causation. It is tempting to reason that since (1) backward causation entails inexplicable coincidences, and since (2) inexplicable coincidences are highly improbable, then (3) backward causation must be highly improbable and therefore should never be postulated. However, I shall argue that this reasoning is unsound because premise 2 is mistaken. More specifically, I shall distinguish between two kinds of inexplicable coincidence: the familiar, admittedly improbable, kind, which I'll call 'Humean", since it involves the type of causation characterized by Hume (discussed in chapter 8), and an unfamiliar, conceptually strange kind of coincidence, which I'll call "non-Humean", since it hinges on a radical change in our usual idea of causation. After articulating this distinction, I shall return to the case of the dancing chief and argue that certain possible results of the bilking experiment are inexplicable coincidences of the unfamiliar, non-Humean sort—the sort that is *not* improbable. Thus it turns out that the force of the bilking argument, at least in certain cases, is not that the backwards causation hypothesis is empirically improbable, but that its postulation involves various conceptually drastic consequences.

Let us consider, to begin with, the two forms of uncaused correlation. In chapter 4, after the discussion of entropy, we encountered a related, and equally pervasive, temporally asymmetric feature of the universe: the fork asymmetry. This may be described as the fact that highly correlated events are always constitutents of a V-shaped pattern of correlation. Very roughly speaking, if events of type A and B are associated with one another, then either there is always a chain of events between them (a chain linked by 'basic nomological determination', as described in chapter 8), or else we find an earlier event of type C that links up with A and B by two such chains of events. What we do not see is the pattern depicted on the right of figure

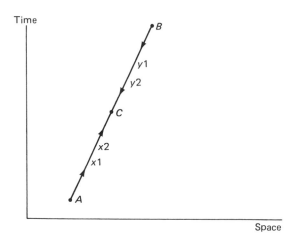

Figure 20

19—an inverse fork—in which A and B are connected only with a characteristic subsequent event, but no preceding one.

The fork asymmetry is usually characterized in causal terms, as implying that correlated events are always causally connected. This formulation follows from the principle of V-correlation, if we combine it with a Humean conception of causation: roughly, that a cause is an earlier member of a chain of direct nomological determination. However, since we are entertaining the possibility of giving up a vital element of the Humean conception, it is important to see how the empirical phenomenon of V-correlation could still be captured. Thus we have

$$\text{Principle of} \atop \text{causal correlation} = {\text{Principle of} \atop \text{V-correlation}} + {\text{Humean conception} \atop \text{of causation}}$$

This analysis allows us to distinguish two ways in which the principle of causal correlation may be violated: (1) by a pattern of events that violates V-correlation and (2) by an abandonment of the Humean conception of causation. The first type of violation could be described as a pattern of correlated events with no *Humean* causal connection between them. It is exhibited in figure 19 on the right. The second type of violation is exemplified in the following rather artificial case, depicted in figure 20. Let A, C, and B be the positions occupied by any particle of a certain sort that is given a particular velocity and then left alone. Now imagine the following extremely peculiar description of this phenomenon. Suppose we say that, among these correlated events, causation works toward the future before C and toward the past after C. That is, A causes $x1$, which causes $x2$, which causes C. And B causes $y1$, which causes $y2$, which causes C. This bizarre account would involve an uncaused correlation between A and B, but there would be no violation of V-correlation.

My point is that one can concoct hypothetical cases of uncaused correlation that do not conform to the familiar idea of 'unlikely coincidence'. Thus I am distinguishing between 'Humean' and 'non-Humean' uncaused correlations. The former, familiar variety violate the principle of V-correlation and are therefore empirically improbable. The latter kind arise from an abandonment of Hume's conception of causation, and however theoretically unacceptable they may be, they cannot be described as unlikely or infrequent. Theories involving non-Humean uncaused correlation may be conceptually unattractive but not empirically improbable.

4. What is shown by the bilking argument?

Let me now apply these points to the bilking argument. We saw that its conclusion was that backward causation implies the existence of inexplicable correlations. But what is the import of this result? Can we infer from it that backward causation is impossible, definitely nonexistent, improbable, or what?

In the first place, there is certainly no contradiction or incoherence in admitting correlations that are not causally connected. Consequently the bilking argument cannot be taken to prove that backward causation is conceptually impossible. Second, we have just seen that not all types of inexplicable correlation constitute improbable coincidences. Therefore whether we can conclude from the bilking argument that backward causation is improbable depends on what kinds of violation of the principle of causal correlation would be associated with the experiment. No doubt it is possible to imagine results of the bilking thought experiment that are indeed improbable. For some conceivable results involve coincidences that violate the principle of V-correlation. But, as we have just seen, not all cases of uncaused correlation fall into this category: for example, the non-Humean free particle depicted just now. Indeed, some of the bilking experiment results may be construed as uncaused correlations of just this sort. And since no violation of V-correlation is involved in these cases, they cannot be regarded as improbable. Let me try to illustrate this idea for the case of Dummett's dancing chief. The point will emerge with greater clarity in the next section, when we examine more realistic physical phenomena.

Suppose that the chief regularly dances after finding that the men were cowardly. Given that dancing has normally caused bravery, it is strange that the chief should lose his retroactive power just in the context of the bilking experiment. However, this correlation between (1) the presence of bilking conditions and (2) the absence of circumstances C needed for dancing to cause bravery, would not have been inexplicable if we had adopted an orthodox causal interpretation in the first place. For suppose we had said that the bravery of the men, in the absence of bilking conditions, constitutes a mysterious, yet orthodox, hidden cause of the joint occurrence of dancing and conditions C. And suppose we had said, in addition, that a variation of the cause—specifically, cowardice plus bilking conditions—has a different effect, namely, dancing plus the absence of C. Then, although there is still a question about what the operative laws of nature could possibly be, there is no inexplicable coincidence in the offing—the correlation between bilking and the chief's loss of power is potentially explainable. The fact that we can give an orthodox causal interpretation of the phenomena without having to recognize uncaused correlations shows that no violation of V-correlation is involved and that the uncaused correlation that emerged in the backward-causation interpretation is of the non-Humean variety.

Of course there is no guarantee that we can always find an orthodox causal interpretation that does not violate the principle of V-correlation. It might be that even on close investigation we can discover no chain of events linking the earlier event (bravery) with antecedents of the later event (inclination to dance). In that case a failure of V-correlation is present, regardless of whether we perform the bilking experiment. If the bilking experiment is then performed, the result might be that every time cowardice is reported, something happens, by pure chance, to prevent the chief from attempting to dance. Once, he is bitten by a snake, then his village is attacked by a neighboring tribe, and so on. This sort of outcome is indeed inexplicable and unlikely. And it implies that the particular type of backward causation that would give rise to it is unlikely.

Thus the import of the bilking argument is slight. It underscores the fact that backward causation goes hand in hand with uncaused correlations. However, the significance of this result as an objection to backward causation is diminished by two considerations. In the first place, there are types of uncaused correlation that are not improbable, and certain forms of backward causation would engender precisely that type of uncaused correlation. The bilking argument does not imply that such forms of backward causation are improbable. Instead, it implies a further revision in our conception of causation: namely, an abandonment of the principle that correlated events are causally connected.

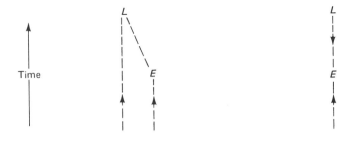

Figure 21

In the second place, we should consider whether or not the bilking argument reveals anything that is not already evident. In other words, do we really need the bilking thought experiment in order to extract from backward causation the potential for uncaused correlations? Are not inexplicable coincidences present, even in the absence of deliberate bilking? It seems to me that the bilking argument is indeed redundant. It merely *highlights* the fact that backward causation involves uncaused correlations, and does not reveal it, because this fact could perfectly well have been discerned without reference to bilking considerations. To see this, consider whether or not there are antecedent conditions that determine the occurrence of the later event L, but without passing through the earlier event E. If there are such conditions, then that is to acknowledge that there is a pattern of events (pictured on the left of figure 21) that violates the principle of V-correlation. Such a form of backward causation can be judged improbable without having to rely on the bilking argument. On the other hand, if the antecedent conditions that determine L do so only via E, then we have a pattern of events analogous to our example of non-Humean, uncaused correlation. Again the uncaused correlation —this time of the sort that does not violate V-correlation—is evident independently of bilking considerations. Thus it seems that the bilking argument is not only weak but unnecessary.

5. Tachyons and positrons

There is no need to rely on far-fetched stories to make these points. Here are a couple of realistic examples from physics.

It has been speculated that there might exist particles—to be called "tachyons"—that go faster than light. Against this hypothesis, however, one often hears the following counterargument. Any superluminal particle would entail backward causation; but backward causation is ruled out by bilking considerations; therefore tachyons do not exist. The import of my preceding discussion is that this argu-

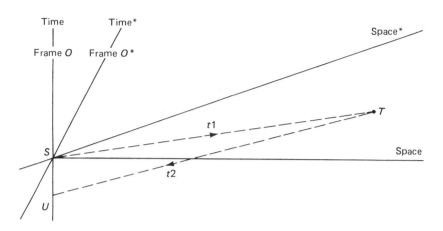

Figure 22
Note that the acuteness of the angle between $O\star$'s space and time dimensions represents the velocity of $O\star$ relative to O. It enables us to see how events that are simultaneous in O (i.e., the line between them would be parallel to O's space dimension) are not simultaneous in $O\star$.

ment against tachyons will not work. Let me repeat that line of thought in the context of this example.

If a particle, $t1$, travels faster than light between spacetime points S and T, then there are inertial frames of reference in which $t1$ is at S before T, and there are other frames in which it is at T before S. This result comes from the special theory of relativity. Now, if we suppose that $t1$'s direction is *objectively* from S to T—that $t1$ goes from S to T in all frames—then it follows that in some frames (e.g., $O\star$) its arrival at T occurs *earlier* than its departure from S.

So far so good. However, things could be arranged so that as soon as the tachyon $t1$ arrives at T, another one $t2$ is instantly sent back from T to $t1$'s place of origin. Moreover this could be done in such a way that $t2$ arrives at that origin at a time before the time at which $t1$ was transmitted, as shown in figure 22. (See Earman 1972 and Redhead 1983 for a fuller account of this setup.)

This situation provides the opportunity for a bilking experiment. Suppose we attach to our tachyon-firing apparatus, O, a bomb/sensor device, which will detect the presence of any tachyon in the immediate vicinity and instantly cause an explosion that will prevent any subsequent firing of tachyons. And suppose the apparatus is primed so that a tachyon will be fired from S unless prevented by such an explosion. Now, it may seem that we are in trouble. For if $t1$ is fired, then $t2$ is fired, and then there is an explosion, and so $t1$ is not

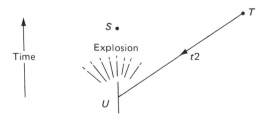

Figure 23

fired. But if *t*1 is not fired, then *t*2 is not fired, so there is no explosion, and so *t*1 is fired. We have a contradiction. Therefore there can be no such setup. Therefore tachyons do not exist.

So the counterargument goes. But, just as in the case of Dummett's dancing chief, it is unsound. One perfectly possible result of the setup is depicted in figure 23–namely, the bomb is exploded by a tachyon arriving from *T* even though, because of the explosion, no tachyon is subsequently sent from *S* to *T*. This outcome involves an inexplicable coincidence. There is an uncaused correlation between making the bilking modification of the original apparatus and the timely transmission of a tachyon from *T*, unprovoked by the arrival of *t*1 from *S*.

However, it would be a mistake to assimilate this uncaused correlation to the familiar Humean variety, and thereby to conclude that this result is unlikely. To see this, note that the violation of the principle of causal correlation would have been avoided if we had given an orthodox (future-oriented) causal interpretation of the initial data. Suppose we had never accepted that the tachyon *t*2 is sent from *T* to *U* but had said instead that the worldline between *T* and *U* is in fact that of a tachyon going from *U* to *T*. In that case it is not surprising that if the original conditions are modified by attaching the bomb/ sensor to *O*, then the effect will be to bring about the events depicted in figure 23 but construed now as a particle being fired from *U*, thereby setting off the bomb and preventing the subsequent firing from *S* to *T*. Let me stress that the point of sketching this reinterpretation is not to *recommend* it over the backward-causation hypothesis but rather to show that the phenomena do not exhibit the sort of inexplicable coincidence that would provide the basis for an argument against the plausibility of tachyons.

The bilking argument does not show that tachyons are impossible. Nor does it provide any empirical evidence against them. What it shows is that if we regard their direction of motion as objective, then we are compelled to accept backward causation, and thereby to accept

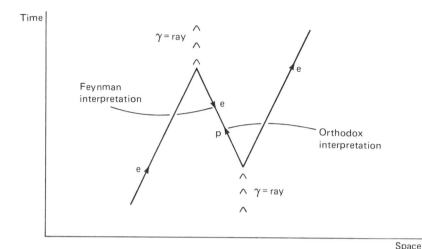

Figure 24

violations of the principle of causal correlation. It should be reiterated that these violations are not of the sort that may be regarded as unlikely. The coincidences that are improbable are cases in which separated events are highly correlated and yet not spatiotemporally connected by a chain of determination: that is to say, violations of the principle of V-correlation. The existence of tachyons engenders un-caused correlations of the second, non-Humean kind, which are not empirically improbable.

Another example from physics, illustrating the weakness of the bilking argument, is the Feynman (1949) theory of positrons, according to which positrons are nothing but electrons moving back-ward in time. An orthodox account of the scenario depicted in figure 24 would say that a high energy gamma ray creates a pair of particles —an electron and positron—that flies off in opposite directions until the positron is anihilated as it collides with another electron. The Feynmann interpretation, on the other hand, involves just one particle—a single electron—that zigzags backward and forward in time as it emits and absorbs gamma rays.

Now consider a bilking experiment. Suppose that things are arranged so that the collision of the backwardly moving electron with a gamma ray instantly activates a shield that prevents any forwardly moving electrons from entering the region (as in figure 25). Thus the cause of collision is put in jeopardy by effects of the collision. What would happen? Given an orthodox construal of the initial setup, the

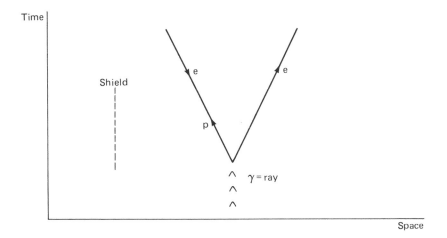

Figure 25

modification would mean that the positron is not anihilated, at least until it leaves the vicinity.

Thus, now construing the situation according to Feynman, we must say that the bilking modification correlates with the lucky arrival of a backwardly moving electron from elsewhere. Again, we see that the possibility of a bilking experiment does not show that Feynman's account is improbable. For the same pattern of events occurs whichever causal interpretation is adopted, and does not involve any violation of V-correlation. If we decide to reject Feynman's theory, this will not reflect a low assessment of its empirical probability. Our basis for preference is, rather, that the restriction to future-oriented causation allows us to preserve two entrenched principles of causal analysis: causes precede their effects, and correlated events are causally connected.

We have reached three main conclusions about the bilking thought experiment. First, it would not automatically falsify our backward-causation hypothesis. Rather, it would confront us with correlations that cannot be causally explained. Second, the prospect of this outcome does not make our hypothesis improbable. Such uncaused correlations can stem directly from our abandonment of the Humean concept of causation rather than from any pattern of events that is empirically unlikely. And third, the association between backward causation and uncaused correlation may be discerned independently of bilking considerations.

6. Empirical evidence against backward causation

It has been useful to distinguish two ways in which causal hypotheses may be evaluated. First, one can abstract away from the particular causal interpretation that is being offered, and focus on the nomological (correlational) structure of events that is implicitly postulated —the relations of necessary and sufficient conditions. And one may assess the empirical likelyhood of such a correlational structure. Second, one may then consider how such a structure—whatever its probability—would best be characterized in causal terms. An interpretation might then be assessed as "incoherent," "somewhat awkward," "perfectly natural," or "clearly correct"—referring now not to the pattern of phenomena that underlies the causal hypothesis but merely to a particular causal description of those phenomena.

Thus, in assessing any specific backward-causation hypothesis, one can separate these two dimensions of evaluation: empirical probability, which is determined by the correlational structure involved; and conceptual desirability, which is determined by how our backward-causation hypothesis fares, compared to other causal interpretations, in conforming with the usual canons of causal discourse. In order to qualify as a clear, unproblematic case of backward causation, it will be necessary to score well on both dimensions—that is, to be a hypothesis implying a correlational structure that is both likely to occur and naturally interpretable in terms of backward causation. And this dual condition explains, I think, why such hypotheses are so few and far between. For there is a tendency for the two desiderata to conflict with one another.

On the one hand, there are phenomena, like those we have just been discussing, that are not empirically improbable. However, we have seen, with the help of the bilking argument, that there is nevertheless a conceptual price to pay for any backward-causation interpretation of such phenomena. An entrenched commitment, beyond the principle that causes precede effects, would have to be given up— namely, the principle that correlated events are causally connected.

On the other hand, there are the phenomena I want to focus on now—phenomena that *are* naturally described in terms of backward causation. These, as we shall see, tend to have the converse problem: the correlational structures they entail are unlikely to occur. In other words, the sort of correlational data that would suggest backward causation are empirically improbable. Let me try, first, to describe in general terms what kinds of data these are that would tend to provoke backward-causation hypotheses; second, to explain why such data do not often occur; and third, to say whether we can ever expect any exceptions. Could there be data that would, on the one hand, suggest backward causation but, on the other hand, not be of the sort whose occurrence we can confidently rule out.

Not surprisingly, the circumstances that will most readily incline us toward the postulation of backward causation will be circumstances in which some temporal pattern that we associate with orthodox causation occurs in the reverse of its usual order. Thus, if events of type A have always caused (and preceded) events of type B, then the observation of B before A might well have some weak tendency to suggest backward causation. A stronger inclination arises when the pattern that is violated (or imagined to be violated) has greater generality. For when the association between causation and the pattern is widespread, it becomes more tempting to connect the pattern with the essence of causation. For example, suppose the entropy of A is lower than the entropy of B, or that A is a free choice and B is a macroscopic event for which A is necessary and sufficient, or that B involves a pair of separated correlated events that are connected by a determination fork to A. In any of these cases the discovery (or supposition) that B precedes A will have a tendency to seem like a case of backward causation. This is part of the reason that Dummett's dancing chief believes in backward causation. And it is the basis of nearly all the unscientific hypothetical scenarios that are discussed in the philosophical literature.

Another circumstance that would strongly suggest backward causation is also present in Dummett's example, and was alluded to at the beginning of this chapter. Suppose a continuous chain of mutually determining events stretches between the distant past and the distant future but that between two nearby events A and B in the chain, it zigzags in time. That is to say, if we imagine ourselves 'moving' along the chain, starting in the past; then when we reach A, we will have to go backward until we get to B. This image reflects entrenched principles of causal interpretation, and it suggests that even though B is earlier than A we will want to say that A causes B. Notice that the zigzag in time involves an inverse fork that violates the principle of V-correlation. Thus this case is like the circumstances described previously, in requiring the temporal mirror image of a familiar pattern of events.

But now the question, "Why do we not observe the kinds of phenomena that would tempt us to invoke backward causation?" has an obvious answer. Any such phenomenon would constitute the time reverse of a process that is de facto irreversible. That is to say, the most striking evidence for backward causation would have to be purely imaginary. For it would have to involve an event of type A

following one of type *B*, where it is a consequence of principles of extreme generality that *A* always precedes *B*. More specifically, the reasons that we do not see the sorts of things that are regarded in the philosophical literature as candidates for backward causation are the reasons, whatever they may be, that entropy increases, that choices precede the events they determine, and that there are no inverse forks.

This by no means implies that there will arise *no* circumstances in which it will be plausible to postulate backward causation. The tachyon and positron cases provide examples in which theoretical economy may arguably be promoted by admitting backward causation, but without any improbable time reverse of familiar structures.

7. Nonconventional predetermination

I began this chapter by describing the most common objection to backward causation: namely, that future directedness is built in, by conventional stipulation, to our concept of causation. This Humean predetermination account was seen to be unsatisfactory in light of Dummett's dancing chief and other hypothetical scenarios that would incline us to postulate backward causation. Therefore, in search of a better strategy, we turned our attention to the bilking argument. Returning now to the first line of thought, notice that our unhappiness with it presupposed that the *pre*determination analysis of causation was intended to be purely a priori. For only given that assumption is it odd for there to be empirical data that would motivate a rejection of the analysis.

Therefore one might still defend a predetermination theory of causation, as long as it is not put forward as an a priori analysis: that is to say, as long as it is acknowledged that our conception of the causal relation has evolved a posteriori. Thus one could maintain that a cluster of central beliefs about causation, modified by experience, has led us to the idea that causation is a complex relation composed of a time-symmetric relation of nomological determination and time order. In that case the future orientation of causation would be an *a posteriori* element of our conceptual scheme. Time order would be an *a posteriori* constitutent of causation.

Our initial objection has no force against this version of the predetermination theory. For even if it is a deeply entrenched belief that causes precede their effects, we can easily recognize hypothetical circumstances that would tempt us to question it. As we have just seen, all that is required is that something *seem* to be a case of causation, yet be oriented backward in time. Even a very 'a priori looking' truth of the form

All A's are B's   (e.g., All blackbirds are black)

may come to seem false if one is convinced that

All $A^\star$'s are A's   (All birds with U-shaped beaks are blackbirds)

and then one observes an $A^\star$ that is not B. Thus, to devise hypothetical evidence for backward causation, it suffices to see how the relation between an event L and an earlier event E can exhibit nontemporal features that are symptomatic of L's causing E. In other words, we can reconcile the thesis that causes are in fact earlier than their effects, with the existence of hypothetical scenarios suggestive of backward causation, providing there exist fairly reliable nontemporal indicators of causation. The reason why it is impossible to imagine evidence for a week in which Thursday comes before Wednesday is that *nothing* marks a day as Thursday other than its temporal position. But the situation is quite dissimilar in the case of causation. For there are properties other than time order that help us to distinguish a cause from its effects.

Thus a nonconventional form of the predetermination theory would not be susceptible to our initial criticism. However, even if such a theory is correct, any claim to have thereby obtained a complete demonstration and explanation of the nonexistence of backward causation would be an exaggeration. For, in the first place, since it is not a matter of pure *a priori* convention that causes precede effects, backwards causation is not inconceivable. So it remains to be explained *why* there are no generally recognized cases of backward causation. Why do we suppose that the predetermination theory is true, or almost true? What is it about the world, or about ourselves, that has led us to a conception of causation as predominantly future oriented? These questions will be addressed in chapter 8. Second, one may reasonably demand of an adequate demonstration of the direction of causation that it persuade us that evidential circumstances will never arise that will lead us to a new account of the causal relation's temporal structure. In this connection it is relevant to cite the results of the last section. Namely, the phenomena that would most obviously suggest backward causation would be the time reverse of well-known irreversible processes. And finally, a genuine *explanation* of our disbelief in backward causation could be expected to give the basic facts that engender the absence of such evidential circumstances. This might be done along the lines indicated in chapter 4, by reference to the cosmological conditions responsible for de facto irreversibility. In the end we are left with an identification of the contingent and conceptual factors that explain why backward causation is not often postulated and that thereby promote the predetermination theory.

7

Time Travel

---

1. How to do it

It has been claimed by Kurt Gödel (1949b) that time travel is, in some sense, physically possible. Not just that travel into the future is possible—this is a well-known consequence of the Special Theory of Relativity, and of comparatively little philosophical interest. It may be accomplished simply by moving quickly and therefore aging slowly. But also, more controversially, that travel backward in time is possible. Gödel's ground for this provocative view is his discovery of certain solutions of the field equations of General Relativity that permit the existence of closed causal chains. A journey back in time would be nothing more than the "backward" part of such a chain. Thus he is led to the following startling result:

> by making a round trip on a rocket ship it is possible in these worlds [i.e., worlds in which his field equation solutions describe the structure of spacetime] to travel into any region of the past, present, and future and back again, exactly as it is possible in other worlds to travel to distant parts of space. (1949b, p. 560)

Many people would like to deny this claim. Some would quarrel with the physics involved. They might argue that Gödel's solutions are physically incompatible with various known facts about the universe. Others take the position that time travel, especially into the past, is conceptually absurd and a fortiori physically impossible. They might rule out Gödel's solutions in the way that we often reject unacceptable mathematical solutions to physical problems. (For example, using the equation, "distance (in feet) = 16 × time (in seconds) squared," to find out how long a stone would take to fall, say, 64 ft, we obtain $t^2 = 4$, and one of the solutions, minus 2 secs, is dismissed out of hand.)

My aim in this chapter is to defend Gödel's claim against the objection that time travel, as he envisions it, cannot occur since it would engender anomalous consequences. I will consider four alleged para-

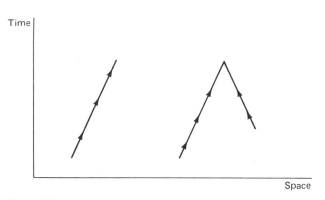

Figure 26

doxes that have been held to refute the possibility of time travel into the past. All but one of them can, I think, be dealt with fairly quickly. So I shall devote most of the chapter to a consideration of the fourth problem—Couldn't I travel back and kill my infant self?—which was noted by Gödel himself and given a powerful formulation by Earman (1972). Evidently this last paradox is a version of the sort of 'bilking' objection to backwards causation that we examined in chapter 6. Since time travel into the past involves backward causation, it should not be surprising that similar considerations would crop up. What is, perhaps, rather striking is that they should turn out to have greater force in this context than they did before.

I shall not attempt to argue for the view that time travel might be achieved by 'jumping' in spacetime (that the worldline of the time traveler may be discontinuous). This project would seem to be both less interesting, since there is no physical theory to give it credence, and more difficult, since extra problems to do with personal identity are involved. In Gödelian time travel a trip back in time would be located along the 'backward' part of a closed timelike line. (A timelike line is a spacetime path along which it is physically possible that there be a causal chain). There would be no particular point at which the traveler would begin the backward phase of his journey. Rather, his worldline would always be oriented toward the locally defined future, yet the curved global structure of spacetime—the way the local regions are pieced together—would permit his arrival home to precede his departure.

One of the reasons that this mode of time travel can be hard to imagine is that one tends to think in terms of 'flat' Euclidean spacetime, represented in figure 26, in which the endpoint of a future-oriented worldline must be later than its origin, and in which an

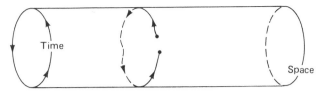

Figure 27

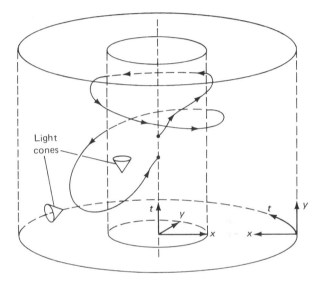

Figure 28

inverted V-shaped worldline is naturally construed, not as a single particle beginning to move backward in time, but as two particles that collide and destroy each other.

In order to loosen the imagination, it can be helpful to think instead of a cylinder-surface spacetime, whose time dimension is a circle (as in figure 27). Here it is evidently quite possible for the endpoint of a particle to be locally earlier than its origin.

Gödel's solution, however, are more complicated than this. A slightly better idea of how time travel would take place in a Gödelean world is conveyed in figure 28. For the sake of picturability, there are only two space dimensions. The spacetime has two parts. There is an internal cylinder whose light cones (within which all possible world-lines must lie) are parallel to the central vertical axis. The rest of the spacetime surrounds the cylinder. In that outer part the time dimension is circular, threading the light cones which are perpendicular to the central axis. Gödel's best-known solution (1949a) is analogous to this spacetime; except, instead of the disjoint internal and external parts that I have depicted, his solution involves a gradual transition between the two extremes. David Malament (1985a) provides an illuminating description and discussion of this model.

Now, consider a worldline that begins in the central axis, moves upward and outward until it reaches the cylinder boundary, then spirals down outside the cylinder, reenters it, and moves upward and inward to the center again. The endpoint of this line is earlier than its origin, although at every point (neglecting the cylinder boundary) it is oriented toward the future.

Let us turn now to the alleged paradoxes of time travel.

2. Is "time travel" an oxymoron?

To travel in time is to traverse some temporal interval in a time that differs from the duration of that interval. Thus we have a straightforward contradiction. Donald Williams (1951) uses a version of this paradox to fault H. G. Wells.

But the problem is instantly solved once it is noted that the magnitude of the temporal interval that is traversed and the duration of the journey are measured in different frames of reference. Suppose I could go to the year 2500 in my time machine. Then it might sound absurd to claim that I could traverse 500 years in 15 minutes. However, the apparent contradiction is resolved once we see that the 500 years refers to Earth time, whereas the 15 minutes refers to an interval measured by clocks in the time machine. Giving up the notion of absolute time and relativizing time to frames of reference allows us to view the 500 years of Earth time and the 15 minutes of the proper time of the time machine as equally good and correct measures of the temporal difference between the year 2500 and the departure of the time machine.

3. Leibniz's law

Unrestricted time travel is incompatible with Leibniz's law (that identical objects have all the same properties). For suppose Charles, who was clean-shaven in 1960, has by 1970 grown a beard, and then travels back to 1960. The early Charles, Charles I, is the same person as the time traveler, Charles II. Therefore, according to Leibniz's law, Charles I and Charles II should have the same properties. Yet Charles II is bearded, whereas Charles I is not.

Some people have thought that cases such as this create difficulties for Leibniz's law even when no time travel is involved. For suppose that Charles does not return to 1960 but continues to live a normal life. One might still argue that Charles I, being the same person as Charles II, should, according to Leibniz's law, have the same properties. To this we can answer that insufficient attention has been paid to tenses. Insofar as the property, having a beard, is used timelessly to mean having a beard at some time or other, then it is true both of Charles I and of Charles II that they exemplify the property. Alternatively, if the property is used with a particular temporal index built in, such as having a beard in 1960, then it is true of both or false of both. When it is said that Charles I, unlike Charles II, has no beard, the temporal indexes are suppressed because they are indicated by the choice of names. But if we are more explicit, we would say that Charles I has no beard in 1960 and Charles II has a beard in 1970. And this is harmless, since it is not the case that one and the same property is both affirmed of Charles I and denied of Charles II.

Does this solution work in our original case in which Charles II travels back to 1960? On the face of it, it does not, since here we do want to claim that Charles I is clean-shaven in 1960 and that Charles II has a beard in 1960. Thus it appears that we violate Leibniz's law in ascribing contradictory properties at the same time to a single individual. But there is an obvious strategy for avoiding this problem. We perform the same sort of maneuver we used earlier when we insisted that the temporal index of the property be made explicit. In this case we must insist that the Charles-proper-time index of the property be made explicit. This is defined by a clock that Charles carries with him throughout his life, and it applies only to events that take place close to him. Similarly, the Earth times, 1960, 1970, and so on, apply straightforwardly, in a Gödelian universe, only to events that occur on, or near, the Earth (although it is possible to extend their application to distant events, in ways that are to some extent conventional). Now, if the property, having a beard in 1960, is used proper timelessly to mean having a beard in 1960 at some Charles proper time or other, then it is true of Charles I and of Charles II that they exemplify it. Also, if some proper-time index is built into the property, then it is either true of both or false of both. When it is said that Charles I does have a beard in 1960 but Charles II does not, the proper-time index is suppressed. But were we more explicit, we would say that Charles I has no beard in 1960 at some proper time t, whereas Charles II has a beard in 1960 at some different proper time t'. As before we have no reason to deny that Charles I is Charles II, and Leibniz's law remains perfectly satisfied.

4. Changing the past

Whatever has already happened cannot now be undone. But if someone could return to some time in the past, he would be able to bring
about a state of affairs that, as a matter of historical fact, did not occur
at that time. This is a contradiction. Therefore we can infer either
than such time travel is impossible or that time travelers are constrained by mysterious forces that conspire to prevent them from
bringing about such contradictions. In particular, suppose that
Charles was not present at the Battle of Hastings. If he could travel
back in time, he could return to 1066, and unless his freedom were
impaired in an unusual manner, he could undo the past and bring
about a contradiction.

This argument, and arguments like it, are invalid. To see this, it is
crucial to recognize the distinction between *changing* the past and
influencing the past. Time travel would allow one to influence the
past but not to change it. Changing the past is indeed logically impossible. However, no such contradictions are involved in the idea of
influencing the past. There is nothing tricky about this. The same
considerations apply to the future. Here again, we have influence, but
we certainly cannot bring about an event that will not occur.

From the assumption that Charles was not at the battle, it does not
follow that he could not have been there. However, this inference is
required if the argument against time travel is to go through. We may
grant that the possibility of time travel implies that Charles may return to 1066, and grant further that this in turn implies that he could
attend the battle. But this conclusion is perfectly compatible with the
supposition that he did not attend it.

It is important to see that no bizarre constraints on Charles's freedom are entailed by the supposition that he did not attend the battle.
From the fact that someone did not do something, it does not follow
that he was not free to do it. Consequently, from the fact that Charles
was not at the battle, it cannot be inferred that if he were to travel
back in time, he would not be free to attend it. Laws of logic do not
involve the kind of limitation on our ability to choose and act freely
of which it is appropriate to give causal explanations. In particular,
we are not required to explain why it is that if Charles did not attend
the battle, then when he travels back in time he does not attend it.
This is merely an instance of $p \rightarrow p$.

5. Autofanticide

Immediately after his claim about the possibility of time travel, which
I have already quoted, Gödel says:

This state of affairs seems to imply an absurdity, for it enables one, e.g., to travel back into the near past of those places where he has himself lived. There he would find a person who would be himself at some earlier period of his life. Now he could do something to this person which, by his memory, he knows has not happened to him. (1949b, pp. 560–561)

Gödel takes this case so seriously that he feels forced to argue on independent grounds (because of prohibitive fuel requirements) that such instances of time travel are in practice impossible. If, however, what I have said about such cases was correct, then Gödel's concern might well seem to be misguided. It is simply false that the time traveler could do something to this earlier self that he knows has not happened to him, such as punch him on the nose. Furthermore this possibility is not implied by his confrontation with his earlier self, together with his freedom. What may be implied is that he could punch his earlier self on the nose. But whether or not he does this, there will be no contradiction. If he does it, despite his memory to the contrary, this shows that his memory was faulty. On the other hand, if we suppose his memory is veridical, then as we have seen, this in no way detracts from his ability to have behaved otherwise than he did.

Let me put myself in the position of Gödel's time traveler. I go back three years in time, return to my house, and discover a person whom I take to be an earlier version of myself sitting in the living room. I feel absolutely certain that this person is in fact my earlier self, since he looks just like me and he is doing exactly what I remember I was doing on the date in question. Furthermore I am absolutely confident that within the last ten years or so of my life I have never been punched on the nose. However, despite all this I feel, and presumably am, free to punch this person on the nose. And whether or not I do so, no contradiction will be engendered. If I do not, then all is well with my faculties. If I do punch him, then this means that things were not as I remembered them.

Many opponents of time travel would base their position on a well-known variant of Gödel's argument. They suppose that if time travel were possible, then people would be able to return to the past and murder their infant selves. But this form of suicide is impossible (for only those who fail will ever be in a position even to make the attempt). So it follows that time travel is impossible.

Note the resemblance between this alleged paradox and Gödel's. Each case concerns a type of causal chain that, when located along an ordinary, open timelike curve, creates no difficulties. The problems arise when one imagines the same type of causal chain but laid along a

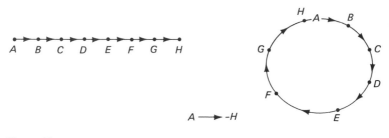

Figure 29

closed timelike curve, as in figure 29. This is because what would normally be the end of the causal chain may be incompatible with what would normally be its beginning. Thus common or garden chains may be self-defeating in the context of a closed timelike curve. No conceptual difficulties are involved in the idea of a causal chain in which someone goes on a journey and kills someone at his destination; or in which someone, who remembers that something has not happened to him, goes on a journey and does that thing to someone else. Problems arise only when we consider these causal chains to be located along closed timelike curves. In this situation it may happen that elements of the chain, which usually would be temporally separated, are now required to coincide with each other; yet their coincidence is impossible. The usual ends of the chains (death of the victim) don't 'fit' with the usual beginnings (good health of the killer).

This account indicates a general question that underlies Gödel's example, the problem of autofanticide, and indeed all the self-defeating-type paradoxes of time travel. Would the existence of closed timelike curves imply the possibility of self-defeating causal chains? Those opponents of time travel who would answer yes, are supposing that any causal chain that may be located along a normal open timelike curve could equally well be located along a closed timelike curve, if any such existed. This view embodies the idea that a timelike curve does not determine some restricted class of causal chains of which it may be the locus. If some causal chain can be located along some timelike line, then it should be locatable along any sort of timelike line—even a closed one, if there is such.

But this idea, as it stands, is surely incorrect. In the first place, it is quite obvious that there are constraints on the timelike curves that may acts as loci for particular sorts of causal chain. The spatiotemporal interval between distinct events of a type of causal process is commonly determined by laws of physics. Second, any causal chain must be consistent with those chains which are located along intersecting

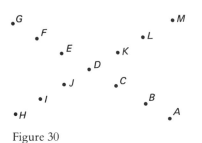

Figure 30

timelike lines. For example, the event D must fit both the chains of which it is an element, as in figure 30. In this respect any causal chain must satisfy consistency conditions imposed by its surroundings.

Therefore it is mistaken to suppose that any causal chain may be located along any sort of timelike curve. The fact that closed causal chains would be subject to consistency conditions is not at all remarkable. It follows that the existence of closed timelike curves and the possibility of time travel do not imply the possibility of self-defeating systems such as Gödel's memory-refuter or the autofanticidal maniac. Such systems are excluded by just those restrictions on causal chains and their interrelationships that ordinarily apply in open time.

6. Bilking time again

The arguments we have just been considering have had the form:

> If time travel were possible, then X would be possible
> But X is impossible
> Therefore time travel is impossible

where X is the implementation of a self-defeating causal chain, such as killing one's infant self or refuting one's memory. Our response to these arguments has been, in each case, to deny the first premise. Thus we concede that autofanticide is impossible but deny that its possibility follows from time travel. My inability to go back in time and kill *myself* as an infant is just a special case of my inability to go back and kill anyone before their death, and this is impossible for the same reason that I cannot, right now, kill someone before his death. The difficulty has nothing to do with time travel.

However, despite the soundness of this response, there should remain a lingering doubt about whether the autofanticide problem, and its variants, have been adequately disposed of. Such misgivings arise, I think, when we focus our attention on cases of *attempted* autofanticide—cases where someone *tries* to realize a self-defeating causal chain. We know of course that success is impossible. Yet repeated failure is nevertheless surprising and disturbing, for it constitutes a violation of well-established regularities. Of course what I have just said is very far from a carefully formulated argument. It does suggest, however, that those whose unhappiness with time travel derives from the consideration of self-defeating causal chains may well not be satisfied with our rebuttal of the original argument. They may contend, rather, that this argument did not successfully capture their objection, and they may cast around for a better way of formulating it.

So far I have been pursuing these issues without any reference to the analysis of backward causation in the last chapter. But time travel into the past is, after all, a form of backward causation, so that discussion is obviously relevant to this one. In particular, there is a clear similarity between the concept of bilking and the idea of a self-defeating causal chain. Consequently we can expect to gain insight into the autofanticide paradox by drawing on our examination of the bilking argument.

In the course of that examination I showed that the bilking considerations can have two possible morals. First, in the case of certain backward-causation hypotheses—for example, the belief of Dummett's dancing chief, the tachyon hypothesis, and Feynmann's theory of positrons—the upshot is merely to underline a certain oddity in causal descriptions of the phenomena, namely, the need to acknowledge uncaused correlations. But in these cases the bilking objection has no tendency to suggest any empirical improbability of the underlying nomological structures. In the second place, however, there are conceivable backward-causation hypotheses whose underlying structures are so-called 'inverse forks'—patterns of phenomena in which an event is nomologically overdetermined by two distinct preceding chains of events. In these cases the bilking argument helps to show that improbable coincidences are required for the correlation between the two branches of the fork to be preserved. We shall see that the case of time travel falls into the second category. Both inverse forks and closed causal chains require a certain lack of randomness in the initial conditions of the universe. In both cases the need for such a constraint is highlighted by bilking considerations, and one can question, on empirical grounds, the likelihood that this constraint is actually satisfied. Let me now make these claims more concrete.

Bilking, in the context of Gödelian time travel, is bringing about some past event that did not occur, such as killing one's infant self or doing something one remembers was not done. Now we know, on the basis of purely semantic reasons, that attempts at such things will invariably fail. But we recognize that there is considerable strangeness in this—something ad hoc and unsatisfying about explaining the repeated failures in terms of changes of mind, guns misfiring, and so forth. Since it is implausible that such mishaps would occur so faithfully over and over again, we conclude instead that there would not be frequent attempts to instantiate self-defeating causal chains. Thus, the existence of time travel implies the nonexistence (or only rare existence) of bilking attempts. However, we have no reason to expect such a fortunate coincidence of time travel with disinterest in bilking. Not only would such an uncaused correlation be an inexplicable coincidence—and highly improbable, given what we see of the world—but also there are facts about human capacities and inclinations that give us positive reason to expect no such correlation. Namely, we know that the ability to travel backward in time would engender attempts at self-defeating causal chains. Consequently we can infer that no such ability will be acquired.

I think that this reasoning is sound. However, its power is to some extent diminished by the fact that it may seem to be all too easy to explain away the repeated failure of bilking attempts, and therefore easy to accommodate the occurrence of bilking attempts. Someone out to kill his early self might get distracted, the gun could jam, or a brilliant surgeon may be on hand to remove the bullet from the infant's brain. And though it seems quite clear how extremely odd it would be for such things to happen over and over again, someone might perhaps still not find it obvious that any incredible violation of rational expectation would be involved. For this reason it is desirable to have a version of the argument that is as free as possible of assumptions about human capacities, and that involves systems whose laws of evolution are deterministic and well understood. A good example of this sort has been devised by Earman:

> . . . consider a rocket ship which at some space-time point x can fire a probe which will travel into the past . . . Suppose that the rocket is programmed to fire the probe unless a safety switch is on and that the safety switch is turned on if and only if the "return" of the probe is detected by a sensing device with which the rocket is equipped. (1972, p. 231)

Thus we face the threat of a self-defeating causal chain. Now, it is clear that continual system failures in the experiment would require highly implausible coincidences. Therefore, it seems that we have no choice but to admit that this type of bilking experiment would not be conducted very often. And so we may infer, as we did earlier, in the autofanticide version of the argument, that time travel into the recent past will not take place on a regular basis.

However, one cannot yet conclude that Gödel's solutions do not describe the structure of our spacetime and that all forms of time travel are ruled out in our world. For the nonexistence of bilking attempts need not be due to the structure of spacetime. Indeed, Gödel argued in just this way that his solutions are not precluded by bilking considerations. To perform a bilking experiment, it would be necessary to travel (or send something) into the *local* past—to some spacetime point that is not only earlier than now but also fairly near here—and it would also be necessary that the (proper) duration of the journey not be longer than the lifetime of the system that is sent back. But these requirements turn out to present insuperable practical difficulties. Specifically, a calculation (e.g., given by Malament 1985b) of the vast amount of fuel required to complete such a trip shows that it would always be technologically impossible to do this (although there are trips into the 'spatially distant past,' for every possible definition of that notion, that require much less fuel and so would not be ruled out). Thus, the nonoccurrence of bilking attempts is explained without having to suppose the nonexistence of closed timelike lines, and therefore provides no evidence for that supposition.

Let me go over this argument more carefully. I am suggesting, in the first place, that we can get an improved version of autofanticide-type arguments if, rather than relying on the premise

If successful bilking is impossible, then time travel is impossible

which we saw was mistaken, we begin instead with the premise

If successful bilking is impossible, then either bilking attempts are rare or circumstances often conspire to ensure that they do not succeed

Of course, successful bilking *is* logically impossible. So we may infer that

Either bilking attempts are not common, or circumstances often conspire to ensure that they fail

But bilking attempts are usually attempts to perform actions that are ordinarily quite easy to perform. Failure is indeed possible for a variety of reasons; but an indefinitely long string of failures, correlated with circumstances leading up to time travel, constitutes a very striking coincidence. Given our experience of the infrequency with which such coincidences occur, we have good reason to believe that the persistent failure of bilking attempts is highly improbable. That is,

It is highly improbable that circumstances will often conspire to ensure that bilking attempts do not succeed

From which, together with the previous statement, we can infer

Bilking attempts are not common

The question now arises as to what accounts for this fact. One might be tempted to say

If bilking attempts are not common, then that is because the structure of spacetime does not permit time travel

and thereby conclude that spacetime is non-Gödelian. However, there are alternative explanations for the rare occurrence of bilking attempts. One alternative is that the initial (de facto) conditions of the universe happen to be arranged in such a way that the individuals who organize trips into the past are not concerned with bilking. This seems unlikely; but it is conceivable. Another possibility is that bilking attempts are precluded, not by the fact that spacetime is non-Gödelian or by a psychological quirk of time travelers, but by some other physical fact that prevents the kind of time travel that would be needed for bilking. Specifically, it may be, as Gödel suggested, that prohibitive fuel requirements preclude convenient time travel into bilking range. Thus, instead of the previous statement, we should say

If bilking attempts are not common, then either spacetime has the wrong structure for time travel, or those who arrange trips to the past do not happen to be interested in bilking, or else the fuel needs for quick trips to the local past are excessive.

Now, the third of these alternatives is quite sufficient to explain why bilking attempts are not common; moreover it is known to be true. Therefore the improbability of bilking attempts gives no reason to suspect that spacetime is non-Gödelian.

This is a rather intricate line of thought, and so I offer one more formulation of it. Perhaps the clearest version of the new bilking argument consists of the following chain of conditionals:

1. If spacetime permits time travel, then men will travel into their local past.
2. If men will travel into their local past, then there will be bilking attempts.
3. Any such bilking attempts will be thwarted.
4. The regular thwarting of bilking attempts will involve an endless string of improbable coincidences.

Therefore
5. If spacetime permits time travel, then there would occur certain phenomena that we have empirical reasons to believe will not in fact occur.

Therefore
6. Spacetime does not permit time travel.

7. A defense of Gödel's position

I have endorsed what I take to be Gödel's response to the preceding argument. Namely, the first premise is incorrect, since too much energy would be need to get back into bilking range. However, the adequacy of this answer is far from uncontroversial. Gödel's position may be challenged from various directions, and I would like to complete this discussion of time travel by attending to four natural criticisms. The first two try to suggest that the bilking objection has no force whatsoever, so there is no need do look for an escape in the practical infeasibility of trips to the recent past. Thus it is supposed that Gödel's response is not needed. The second two complaints make the opposite allegation. They claim that the bilking argument is too strong to be blocked by the mere practical difficulty of travel to the local past. Therefore it is suggested that Gödel's response is powerless.

The first sort of objection to Gödel's position is that any concern with bilking is misguided and unnecessary. One reason that this view might seem plausible is a tendency (e.g., Horwich 1975) to consider only the poor version of the autofanticide argument (described in section 5) and to overlook the possibility of a better formulation.

A second reason that the force of the autofanticide issue may be underestimated is that it may seem doubtful (e.g., Malament 1985a) that the continual failure of bilking attempts would, as I have claimed, require a series of improbable coincidences. After all, it might be said, perhaps all our actions are causally determined by the initial conditions of the universe and the laws of physics. If so, then whenever there are attempts to do things that are contrary to what is determined, those attempts will fail. In any such case circumstances must conspire to make sure that something goes wrong. But we do not suppose that there is anything strange or improbable about such failures. And we certainly don't think that these considerations provide an argument against determinism. So why should we attribute any greater significance to the fact that bilking attempts, in the context of time travel, would regularly be foiled?

The answer is that time travel and determinism are not analogous, because quick time travel into the local past *would* require coincidences whereas determinism would not. If there were a regular practice of travel into the past, then there would have to be a correlation between, for example, *the time traveler having an intention to bump off the child who lives at his old address* and the existence of *circumstances that will frustrate this intention*. And we know enough about human motivation (specifically, about the factors that might produce this bilking inclination) and the kinds of phenomena that could cause this plan to fail (amnesia, gun-jamming, brilliant surgeons, etc.) to claim that any such correlation would be an improbable coincidence. On the other hand, our failure to act in ways that are contrary to what has been determined by past circumstances does not involve any such mysterious coincidence. To help make this clear, let me distinguish between the normal situation in which we *do not* know in advance what we have been determined to do, and the abnormal situation in which we *do* know. In the normal situation there is obviously no positive correlation between the class of intentions to do A (e.g., scratch one's hand) and the class of circumstances that will prevent the performance of A. We can obtain a positive correlation by restricting the first class to "intentions to do A when it is determined that A will not be done." But the correlation between this class and circumstances leading to failure is trivial to explain. Now consider the abnormal situation in which we *do* know what we are determined to do. Suppose someone tries to do A whenever he knows that not-A is determined. In that case, there is a correlation between forming the intention to do A and not succeeding. But again there is no surprise about this correlation. The very condition that determines that act A will not be done *also* stimulates the futile attempt to do it—the correlated events have a common cause. Thus determinism does not involve strange coincidences. Our failure to do what we are determined not to do hardly calls for explanation. On the other hand, time travel into the local past engenders closed causal chains that verge on being self-defeating, and this possibility can be precluded only by means of appropriate coincidences.

A third critical response to Gödel's position is has been given by Earman (1972). He maintains that if there were closed timelike lines, then the mere *physical possibility* of enough fuel for trips into the immediate past would engender paradox, and so he disputes Gödel's claim that their *technological impossibility* provides a way out. I think that this is incorrect. We may grant that there cannot be a physically possible world that contains the conjunction of

1. closed timelike lines
2. sufficient fuel and technology for trips to the local past
3. bilking inclinations
4. initial conditions that preclude coincidences

But when Earman says that the physical possibility of short trips to the past is paradoxical, he has in mind that a counterfactual world in which we hypothetically do have the fuel and technology for such trips would retain certain pertinent characteristics of the actual world: specifically, our spacetime structure, our susceptibility to bilking inclinations, and initial conditions, like ours, that preclude frequent coincidences. He assumes, in other words, that if 1, 3, and 4 are actually the case, then the counterfactual conditional

(E)   If 2 were true, then 1, 2, 3, and 4 would be true

must also be the case—a conditional which implies, falsely, that 2 is impossible. So we are able to conclude that (E) is false, and therefore that either 1, 3, or 4 is false, and the obvious culprit is 1—the existence of closed timelike curves.

However, the inference to the counterfactual is invalid. In a world with closed timelike curves the de facto distribution of matter must develop, according to the laws of motion, in such a way that certain events eventually evolve back into themselves. Therefore it is to be expected that such de facto conditions will be much more severely constrained than in open-time worlds. Consequently we should not be surprised that a hypothetical change in them—such as the supposition of more fuel or better technology—could require a difference of spacetime structure or the sort of initial conditions that will give rise to certain types of coincidence. Thus there certainly are possible worlds, consistent with the laws of nature, in which 1, 3, and 4 obtain and 2 does not, and Earman has given us no reason to suspect that ours is not one of them.

Finally, Gödel's position might be challenged on the grounds that the danger of bilking does not come only from *deliberate* action, so the danger is not removed by the mere *practical* impossibility of intentional bilking stategies. Behind this thought is the idea that, just so long as there are *any* closed timelike lines, there is the prospect of circumstances *naturally* arising that would engender self-defeating causal chains—a prospect that can be blocked only by appropriate initial conditions. In other words, the de facto conditions of a Gödelian universe would have to be so arranged that the events that lie on closed timelike lines eventually cause themselves, with the help of surrounding circumstances. Now, the argument goes, such a happy distribution of circumstances (such an ordered initial state) is highly implausible, given our experience of the actual world. For the observations that have led us to the second law of thermodynamics and to the principle that correlated events are causally connected also suggest that the initial conditions in our world exhibit a certain randomness. And if this is right, then we can infer that the existence of closed timelike curves is unlikely: we are probably not living in a Gödelian universe.

In order to assess this argument, it is useful to recollect the case of cylindrical spacetime, in which any total state must engender itself, via the laws of nature, after a certain period of time. Let us consider whether the data that suggest the second law of thermodynamics give us any reason to suppose that we are not living in such a world? In the first place, those observations do not absolutely rule out that kind of model. As we saw at the end of chapter 4, there is work by Schmidt (1966) and Cocke (1967) suggesting that the familiar tendency of entropy to increase in isolated systems is quite compatible with a cylindrical spacetime. For there are conceivable de facto conditions that would bring about a period of galactic expansion in which the entropies of systems increase, followed by a period of contraction in which entropies decrease, and leading back to precisely the condition from which we 'began'. So there is no conclusive argument from familiar entropic behavior to the thesis that time is not circular. Nevertheless, there might seem to be a probabilistic argument. For the set of very special conditions, in a cylindrical spacetime, that would give rise to the right kind of entropy-fluctuating cycle no doubt comprises a tiny proportion of the set of all the possible initial conditions that would produce the observed entropic facts. And on that basis we might be inclined to conclude that it is very unlikely that those special conditions obtained in our world, and therefore very unlikely that our spacetime is cylindrical. However, there is a big gap in this reasoning. No argument has been given for assuming that all the possible initial conditions of the universe are equally likely. Indeed, it is hard to see any way of establishing the 'right' probability distribution over possible initial conditions, and it seems quite plausible that there is no rational constraint of this sort. Therefore we cannot infer, from the second law data, that our spacetime is probably not cylindrical.

A further question, however, concerns not how probable it is, in light of such experience, that time is circular but rather whether these data provide any evidence against this hypothesis. In other words, do our entropic observations *reduce* the probability that time is circular. Here again, I think the answer, at present, is no. Little is known about what proportion of the possible de facto conditions in cylin-

drical spacetime would yield the entropic behavior that we are familiar with. Consequently we cannot tell if it is more 'difficult' to engender our second law data in a closed world than in an open one.
Therefore we have no reason to count these data as evidence against
circular time.

These points carry over to Gödel's models. Just as in the case of the
cylindrical spacetime, it is necessary that the initial state of a universe
with closed timelike lines possesses a certain order. Moreover those
special initial conditions that will *both* conform to a Gödelian
spacetime *and* engender the entropic behavior we observe, form a
small subset of all the possible initial conditions compatible with our
entropic data. However, in the absence of any uncontroversial way of
attributing a probability distribution to initial conditions, we cannot
infer that the conditions needed to engender the second law data in a
Gödel universe are unlikely. Thus we cannot conclude, in light of
these observations, that there are probably no closed timelike curves.
In addition, it is unclear that the second law data would be engendered by a smaller proportion of Gödelian initial conditions than non-
Gödelian initial conditions. Therefore we cannot even claim that the
second law data reduce the chances that we are in a Gödelian world.

Thus the argument against there being *any* time travel into the past
is much weaker than the bilking argument against time travel into the
local past. In the latter case specific coincidences would ensue, in
whose improbability we can have a great deal of confidence. So we
can conclude that such trips will not take place. However, this fact is
explainable without having to posit a non-Gödelean spacetime.
Therefore, as Gödel said, our own history is indeed out of bounds;
but, for all we know, there do exist closed timelike lines, making a
real possibility of trips to the spatially distant past.

8
Causation

The aim of science may be presented in a variety of ways: to discover laws of nature, to explain phenomena, to identify their causes, to say what would transpire in a range of hypothetical circumstances. None of these characterizations is especially controversial. Everyone will agree that causation, explanation, law of nature, and counterfactual dependence are methodologically vital and intimately related. Disputes arise, however—perhaps the central issues in philosophy of science—as soon as one tries to say precisely what these things are and precisely how they interact with one another.

This clutch of problems will be the general background theme of our next three chapters. More specifically, I will examine the respects in which causation, explanation, and counterfactual dependence are asymmetric in time. But the narrower project cannot be pursued without some attention to broader questions about the nature and affiliation of the concepts involved. Therefore I shall be proposing a rough picture of their internal structure and interrelationships—one that gives a primary role to the concept of explanation and characterizes causation, law, and counterfactual dependence in terms of explanation. My accounts of these notions will, I am afraid, be somewhat crude and not thoroughly defended. The point is to say just enough to sustain my explanations of their temporal properties.

Let me begin with the question of why effects rarely, if ever, precede their causes. This question is not the same as the problem of backward causation, discussed in chapter 6. There we took for granted that the typical direction of causation is toward the future, and asked whether there are any actual or possible deviations from this norm (a cause occurring later than its effect). The answer was no and yes—there aren't any actual cases as far as we know, but their existence is perfectly conceivable. The matter before us now is, rather, why the *predominant* direction of causation goes from past to future. This problem would remain even if there were occasional cases of backward causation.

The reason that the direction of causation is puzzling, and its explanation controversial, is this. On the one hand, causation appears to be a type of *determination*; on the other hand, all the types of determination with which we might plausibly identify causation seem to be time-symmetric. Consider, for example, the relation, "*C* is part of a condition that, given our laws of nature, requires the occurrence of *E*". This holds when *C* is later than *E* just as often as when *C* is earlier than *E*. And apparently the same thing goes for other prospective relations of determination—such as necessity, contiguity, or probabilistic connection—no matter how strong or weak they may be, and no matter how they are combined.

The natural—and, I think, correct—response to this situation, is to conclude that causation shouldn't be identified merely with some species of determination but with determination plus something else. A further ingredient must be added that will explain why causes tend to be earlier than their effects. But then the question arises as to what the extra condition should be, and this is the locus of much of the controversy surrounding the problem. Should we perhaps simply add the stipulation, "*C* is earlier than *E*"? This is the view I have called "conventional predetermination", since it treats time order as an a priori, analytic constituent of causation. Or should we, instead, add a different element to the definition of "causation"—one that happens, as a matter of contingent fact rather than as a matter of meaning, to constrain the time order of cause and effect. This strategy is realized in various alternative views, which I'll call "substantive" accounts, whereby the concept of causation does not in itself engender a time asymmetry but does so only together with other general facts about the world—facts that, had they been different, could have resulted in an opposite direction of causation, facts such as the fixed and settled character of past events (Mackie 1974), the prevalence of decay processes (Reichenbach 1956; Dummett 1964; Papineau 1985), the alleged temporal asymmetry of counterfactual dependence (Lewis 1979b), the orientation of action (Gasking 1955; von Wright 1971; Healey 1983), and the perception of time order (Mellor 1981).

I am going to argue that each of these two general points of view—conventional predetermination and substantive accounts—contain important aspects of the truth and that the right strategy is to combine them in a certain way. To see, in a preliminary fashion, what I have in mind, it is necessary to attend to the distinction between *explaining a fact* and *explaining why we should believe it*. In particular, we must separate the questions, "Why is it the case that causes typically precede their effects?" and "On what basis do we maintain that causes typically precede their effects?" These problems are not always clearly enough distinguished. (Perhaps this is because their answers have been thought to be the same.) According to the conventional predetermination picture, the *fact* is explained trivially by noting that time order is a constituent of causation, and the belief is explained as a priori. According to each of the substantive accounts, the fact is explained by whatever contingent features of the world are involved in generating the direction of causation from whatever definition of "causation" is assumed. And our belief in the future orientation of causation is derived a posteriori from our knowledge of those contingent features of the world. My account will extract pieces from each of these approaches. I will argue that the theory of conventional predetermination has roughly the right answer to why causes typically precede their effects, but that the substantive views give roughly the right explanation of why we ought to believe this. That is to say, I think the direction of causation is explained by noting that time order is a 'constituent' of the causal relation (in a sense to be described later), but this fact is a posteriori.

In presenting this view of the matter, I am going to distinguish between our *concept* of causation (which I take to be the cluster of all of our 'important' beliefs about causation) and our *theory* of causation (a subset of those beliefs—specifically, the subset that allows an especially simply specification of when the relation obtains). I will begin by proposing a certain neo-Humean theory of causation—a theory according to which it is constitutive of the relation of causation that causes typically precede their effects. Then I will turn to the explanation of why we believe this theory, particularly its temporal component. I will show that it can be derived, a posteriori and in several independent ways, from other elements of our concept of causation. Thus responsibility for our committment to the future orientation of causation is spread among many of the ideas that philosophers have thought were the source of it; there is no need to choose among them. In this way my explanation differs from that given by each individual substantive account.

Much of what I shall say about causation is obtained by beginning with the conventional predetermination view and modifying it in two directions. First, we should not think that every analysis of "*X* causes *Y*" in terms of time must be just the conjunction of some determining relation and the further condition '*X* is earlier than *Y*'. More sophisticated theories are possible in which the *earlier than* relation is involved in a complex way, but which still entail that causation is predominantly future oriented. If such a theory of causation is correct, it will explain the normal direction of causation and yet not be embarassed by the odd case of simultaneous causation or epiphenomena.

Second, any claim to the effect that something is 'true by convention', 'true by definition', or 'a priori' will be subject to Quine's (1951) powerful arguments against the viability of those notions. Therefore we must look for an account that can accommodate the ways in which all our commitments have some a posteriori character. To that end we should appreciate that a predetermination account can perfectly well be construed as an *a posteriori* theory rather than an *a priori* conceptual analysis. True, most proposals have been intended as accounts of the latter sort—as claims about the meaning of the word "cause". However, it is possible instead to offer a predetermination account as an a posteriori analysis of causation—an analysis that is intended to be somewhat analogous in status, for example, to the chemical analysis of water—describing the underlying structure of a phenomenon whose familiar symptoms merely provide a moderately reliable guide to when the stuff is present. Similarly I will suppose that the predetermination theory of causation serves to identify, in a precise and authoritative way, the relation that is the subject of the numerous more or less accurate maxims that constitute our concept of causation.

2. A neo-Humean theory of causation

The most influential proponent of building time into causation was David Hume, who characterized a *cause* as

> . . . an object precedent and contiguous to another, and where all the objects resembling the former are placed in like relations of precedence and contiguity to those objects that resemble the latter. (*Treatise*, Bk. 1, Part III, sec. XIV)

In other words

> C causes E   *if and only if* every event like C immediately precedes an event like E

This theory has been improved in the light of some valid objections. Thus (1) it is plain that cause and effect need not be contiguous. Separated events, such as an explosion and the lighting of a fuse, may perfectly well be causally connected when there exists a causal chain from one to the other. (2) Events of different types might invariably follow one another by mere coincidence, and not in virtue of any causal relation between them: for instance, Mr. Smith's withdrawing $666 from his account, and his being struck by lightning as he leaves the bank. This sort of purely accidental correlation need not be especially unlikely if the events are so narrowly characterized as to be very rarely exemplified. (3) Striking a match sometimes causes it to burst into flames and sometimes doesn't, so being a cause of something does not entail constant conjunction with it. (4) Any two events are like one another in some ways, and not in others. Therefore the notion of similarity (which is needed to define 'like causes' and 'like effects') is vacuous in the absence of any indication of the relevant respects of similarity.

These problems have inspired various neo-Humean ('regularity') theories (e.g., Hempel 1965, p. 349; Popper 1972, p. 91). Here is a fairly promising version:

> A direct cause of some effect is an essential part of an antecedent condition whose intrinsic description entails, via basic laws of nature, that the effect will occur. And causation in general involves a chain of direct causation. That is to say, C causes E if and only if there are some events, $e1, e2, \ldots, eN$, such that C directly causes $e1$, $e1$ directly causes $e2, \ldots$, and eN directly causes E.

Briefly then, one can accommodate objection 1, as Hume actually did, by distinguishing direct and indirect causation and dropping the contiguity requirement (for the latter, at least); objection 2 is set aside by invoking the concept of law, and subsequently addressing (as we shall in chapter 10) the question of which regularities are to count as laws; objection 3 is dealt with by allowing that a mere part of the total cause may qualify as a cause; and as for objection 4, the significant respects of similarity are implicitly specified as intrinsic characteristics relevant to basic laws of nature. We require that the descriptions be *intrinsic* (in the sense discussed in chapter 3) in order to preclude the use of event characterizations that contain information about other events. Otherwise, a cleverly contrived description of C (e.g., a sneeze, described as "the sneeze that occurred twenty seconds before Caesar was stabbed") could, together with laws of nature, entail the occurrence of a causally unrelated event E (Caesar's death), merely by incorporating information about some event (the stabbing) that does cause E.

The preceding discussion is, of course only a sketch. Each move and countermove deserves a fuller treatment than I can give here without straying too far from the question at issue. I mention them only because my account of causal direction is set within the context of a neo-Humean regularity theory, and so it is important to indicate that such an account of causation is a live option. True, there are difficulties—but no compelling reason, I think, to suppose that they can't be overcome.

The account requires that the laws be explanatorily *basic* in order to distinguish direct from indirect cases of nomological determination, and thereby to be in a position to deal with certain standard objections, as we shall see. Our picture of direct determination is roughly as follows. When one type of event, C, nomologically determines another type E, then there is often a third event D such that the determination of E by C is explained by the determination of D by C and the determination of E by D. Such explanatory relations are most clear when the law $L1$ involved in going from C to D is different from the law $L2$ needed to get from D to E. In that case it is obvious that the nomological relation between C and E—namely, $L1$ & $L2$— is explained by the two more basic laws, $L1$ and $L2$. However, even if only a single law is involved in the process, there is still a sense in which the law relating C and E is explained by the law taking us from C to D, and from D to E. For example, suppose there is a law stating that whenever an event of type x occurs at spacetime region r, then an event of type $O(x)$ must occur at spacetime region $F(r)$. Now suppose that C is an event of type x occurring at $r1$, D an event of type $O(x)$ occurring at $F(r1)$, and E an event of type $O(O(x))$ occurring at $F(F(r1))$. Then, although there is some sense of "involvement" in which the same law is involved in going from C to D as in going from C to E, there is another sense in which different laws are involved. For, unlike the case of C and D, the nomological relation of which C and E are *instances*, and in virtue of which C determines E, is the following *derived* law: if an event of type x occurs at r, then an event of type $O(O(x))$ must occur at $F(F(r))$. In general, as we continue to interpolate events between determining events, we encounter relations that are more and more basic. In the limit we find relations of determination by means of the most basic laws, and these we identify with the elementary 'links' in a determination chain. Consider, for example, a law that specifies some aspect of the state of a system as a function of the state's initial condition and the elapsed time

$$St(t) = f(St(0), t)$$

If time were discrete, then we could reduce this law to more basic terms, so that it specified the state at one time in terms of the state at the preceding time

$$St(t + 1) = g(St(t))$$

However, since time is continuous, the most direct cases of determination are those in which the state at a given time determines the state at an infinitesimally different time, and the most basic laws are the differential equations that describe this form of determination:

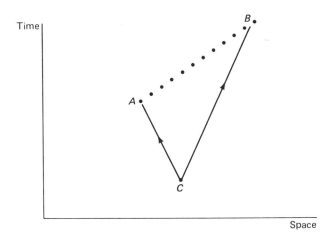

Figure 31

$$St(t + dt) = g(St(t)) \quad \text{or} \quad \frac{dSt}{dt} = h$$

I hope that these sketchy remarks help to make tolerably clear the idea of *direct* determination via *basic* law. Later on, some further light will be shed on 'laws' and what makes them 'explanatorily basic'.

One virtue of defining causation in terms of chains of direct determination is that we are able to avoid confounding cause and effect with mere epiphenomena. The latter are events that are nomologically related, not because one causes the other but through being common effects of the same cause. In such cases an event may determine a later event and yet not cause it. For example, a flash of lightning does not cause the clap of thunder that follows, and a change in barometer reading does not cause the subsequent storm. Schematically, cases in which A and B are epiphenomenal effects of C (as in figure 31) are distinguishable by the fact that the determination of B by A is explained by the existence of relations of determination between some earlier event C, and A, and between C and B. Thus the determination chain between A and B passes through C. So the only way for A to cause B would be for A to cause C and C to cause B. Therefore, given the absence of backward causation, it must be that C causes A and B.

A second advantage of defining causation in terms of chains of direct determination is that it helps us to solve a problem presented by simultaneous causation. The trouble is that, as it stands, the predetermination theory requires that a direct cause be earlier than its effects, and it therefore precludes simultaneous causation. But this is

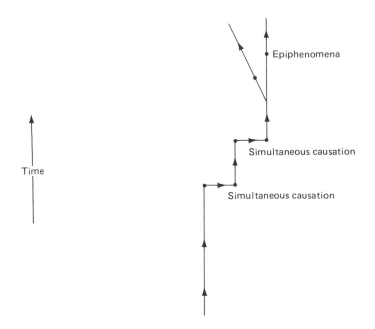

Figure 32

objectionable; for we are perfectly familiar with cases of causes being simultaneous with their effects—pushing or pulling, for example, where the application of a force to an object causes a simultaneous acceleration of that object.

In order to solve this problem, the role of time in the theory must be changed slightly. We should not impose the time-order requirement on each basic link in a causal chain. Instead, we should consider *entire* chains of determination (stretching between the distant past and the distant future) and then impose on the elements of any such chain a causal interpretation subject to the following pair of global contraints: (1) require that causes precede their non-simultaneous effects, and (2) maximize causal continuity (so that causal priority in one part of the chain may be 'smoothly' extended to adjacent parts). In other words, we begin with a chain of direct nomological determination strung out through time and perhaps containing occasional simultaneous links, as in figure 32. We now wish to associate 'arrows', representing the causal relation, with the basic parts of the chain, and in order to do so, we deploy constraints 1 and 2. We find, first, that the continuity condition tells us to avoid causal interpretations in which adjacent links have arrows that point in opposite directions. So that condition severely constrains the number of possible overall causal interpretations of the chain. And this number is then narrowed down to a single possibility when we impose the requirement that causes precede their non-simultaneous effects. Thus time order is invoked to determine the usual direction of causation, in such a way that cases of simultaneous are not automatically precluded.

Backward causation is of course still ruled out. But note that this feature of the theory is by no means unfortunate, given that, as far as we know, there is no such thing. One might well have objected to our theory, were it put forward as an a priori definition. For no doubt backward causation is conceivable. But since the theory purports merely to describe a posteriori the actual nature of causation, it is not threatened by the fact that under different evidential conditions we would be inclined to take a different view of the matter.

I hope to have made it plausible in the last few pages that certain notorious objections to the Humean approach may be met. Thus I am resisting the view, propounded, for example, by Lewis (1973b), that problems arising from epiphenomena and backward causation torpedo the Humean approach and encourage us to look for an alternative account of causation in terms of counterfactual dependence. Not only is there little reason to give up on regularity analyses, but when we examine the counterfactual alternative (in chapter 10), we shall encounter difficulties that are, I think, even more substantial than those we find in the present approach to causation.

One final point—for we shouldn't leave the theory of causation without saying at least something about the question of *probabilistic* causation. Is it really necessary for causes to *determine* their effects? Here I have in mind not the issue of whether what we call "laws" are anything more than extremely widespread regularities but rather whether causes lead *invariably* to their effects. Hume's opinion—that they do, that given the same causes, the same effects will occur—is embodied in our practice of distinguishing causal from statistical laws and in formulating "determinism" as the thesis that every event has a cause. His idea is reflected in our analysis by the requirement that the effect be *deducible* from laws and initial conditions. On the other hand, there are occasions on which we recognize causes that neither determine nor play any role in determining what we regard as their effects (Anscombe 1971). For example, an overexposure to radiation might well be said to cause an illness, even when it is believed that this result was to some degree purely random.

My own view of this matter is that there is a genuine ambiguity at the bottom of it: the word "cause" has two meanings, one involving the notion of invariability and the other not. However, I won't presuppose that this is correct. Rather, I shall try to explicate a weaker conception of causation, and I shall assume that the stronger one may be obtained by adding a requirement of determination. What will matter, for our purposes, is that the weaker explication provide *necessary* conditions for causation that explain the direction of causation. Whether or not those conditions are also sufficient for *the*, or merely for *a*, concept of causation will not be important for the issue at hand.

One way of modifying our analysis to obtain an appropriately weaker conception is to let the circumstances of the alleged cause include not only surrounding initial conditions but also facts about the upshot of random processes. In other words, if the initial conditions, basic probabilistic laws of nature, and specific facts about how, given those initial conditions, the indeterministic factors will turn out together entail the occurrence of the event, then we shall say that those conditions directly caused the event. Thus suppose (rather implausibly) there is a basic statistical law that implies that people exposed to a certain radiation overdose stand a 20 percent chance of contracting leukemia. And suppose Smith gets both the overdose and leukemia. We may conclude (in the absence of any overdetermining factors) that the radiation caused the disease.

I have characterized the structure of causation by starting from a form of nomological connection and adding requirements of causal continuity and time order. Given this theory of what causation is, the explanation of why causes typically precede their effects is simply that this feature is *constitutive* of the causal relation—which is tantamount to saying that the fact has *no* explanation. For, in general, the theory that describes the nature of a type of entity may be deployed to explain further properties of the type, but the theory itself is not subject to explanation. Thus one does not ask why water contains oxygen or why planets orbit starts. Similarly no distinct fact is the explanatory basis for the direction of causation. To this extent my 'explanation' parallels that of the conventional predetermination model. Therefore it is equally susceptible to a certain standard objection, which I would like to discuss next.

3. Manipulation

A common charge is that if the direction of causation is to explain, as it should, the blatant irrationality of acting for the sake of the past, then the direction of causation cannot result simply from the presence of a temporal constituent in the causal relation. More specifically, Dummett (1954), Flew (1954), and Mackie (1974) have felt that if, for example, a cause were nothing more than an *earlier* necessary condition, then it would be arbitrary and wrong to require of a rational act that it be intended to *cause* a desirable event. For there would be no reason why it wouldn't be just as good for one's act to 'retro-cause' a desirable event—where a 'retro-cause' is simply a *later* necessary condition. Therefore the predetermination model allegedly leaves us unable to explain why we do not do things now to ensure the occurrence of past events.

My response to this argument is to deny the presupposition that in order to be rational, an act must be intended to cause, rather than to 'retro-cause', a desirable event. It might be thought that in the absence of some such restriction the door would be wide open to wholesale action for the sake of the past. But that would be a mistake. For, even without the restriction, a past-oriented decision could be motivated only in very exceptional circumstances. To justify this claim, let me sketch a line of thought that will be fully developed in chapter 9. A past-oriented act may be rational only if we believe that it is *needed* to 'retro-cause' some earlier desirable event—that is, only if we are not already aware of some state that will retro-cause the event. And this condition will almost never be satisfied. For the chains of determination between our actions and past events are mediated by our motivational states of belief and desire. Therefore there is usually no potential retro-effect of a prospective action that we cannot *already* know will, or will not, be retro-caused by attention to our beliefs and desires. The only exceptions will occur in very contrived, hypothetical situations, such as Newcomb's decision context. And in such cases it is far from clear that past-oriented action actually is irrational. Thus the principle, "Act only for the sake of what might be caused," isn't needed to explain why we don't, in practice, act for the sake of the past. And in that case we can agree with Dummett, Flew, and Mackie that this principle does not square well with the predetermination-chain theory of causation. But, unlike them, we can regard the tension as grounds for rejecting, not the theory of causation, but the causal principle of rational choice.

The idea that there is an intimate connection between causal priority and rational motivation has been elevated to the status of an analysis of causation in the work of Gasking (1955) and von Wright (1971). Their 'manipulability' proposal is that the content of a causal generalization "*C* causes *E*" is something like "*C* could be used as a means for producing *E*", or "*C* may be directly controlled as a method of indirectly controlling *E*". But what can this possibly mean? If the analysis stops here, we are left with notions, like 'means' and 'method', that are more obscure than the one we were trying to explain. But, as Judith Thomson (1977) has pointed out, if we continue in the natural way, we would have to admit that what the criterion says is "There could be an action W such that W causes C and C causes E"—a construal whose reference to causation evidently disqualifies it as an analysis of that idea. The basic problem seems to be that the manipulability analysis is taking us in the wrong direction. It is more natural to explain the notions 'means' and 'method' in terms of causation rather than the other way round.

This problem is avoided in a formulation discussed by Richard Healey (1983, p. 92):

> . . . of two causally related events X and Y, if a person can produce an event of type Y without producing an event of type X, but not vice versa, then X causes Y.

However, this is tantamount to the view that any sufficient but unnecessary condition is a cause. It suffices to consider cases like the dryness and lighting of a match to see that it will not quite do as it stands. For in some circumstances a match can be dried without lighting it, but not vice versa; yet surely the lighting does not cause the dryness as Healey's criterion would seem to imply.

A better way of understanding the role of manipulability, in my opinion, is to regard it as providing one of the cluster of crude, fallible reference-fixing principles that help us to zero in on the causal relation. Thus we might roughly identify causation as "that relation believed to hold between rational choices and the desired events for the sake of which those choices are made". This maxim picks out a *future*-oriented relation. For, as we have just seen, it is typically irrational (for reasons independent of causation) to act for the sake of the past.

In my view, the manipulability maxim is not strictly accurate. There are occasions when we *should* act for the sake of events that we do *not* hope to cause, as we shall see when we examine Newcomb's problem in chapter 11. However, this departure from the truth shows merely that this maxim cannot be thought to capture the essential nature of causation. But it is not thereby disqualified from inclusion in the cluther of beliefs that makes up our concept of causation. It may perfectly well be a defeasible element of the concept, helping us to pick out a relation of causation whose underlying structure will then be described, in quite different terms, by the predetermination-chain account.

Thus I am supposing that the relationship between the concept and theory of causation is something like the relationship between our concept and theory of other parts of the world. For example, *water* is not constituted and identified by means of a set of analytically necessary and sufficient observable properties. Rather, it is recognized by fallible criteria, such as 'colorless, tasteless liquid' and 'constituent of rain'. Its nature as H_2O is discovered afterward, and is then used to correct the original principles of identification. Similarly in the case of causation we should allow that its structure, as given by the predetermination-chain theory, may conflict with the symptoms by which we recognize instances of causation. These symptoms need be neither essential, universal, nor permanent. In particular, it could be that causation is identified in a loose way by maxims including (M)— "Causation is the relation believed to hold between rational choices and the desired events for the sake of which those choices are made"—that help to point us toward causation but do not describe its nature. The relation so ostended can then turn out to have a structure in which predominant time order is built in. And having arrived at this view of causation, we might then acknowledge cases (e.g., Newcomb's decision context) in which our initial maxims of identification are violated.

4. Why believe that cause precedes effect?

So far I have focused on the *metaphysical* question of why causes typically precede their effects. And I have answered it by suggesting a theory of causation in which time order is a constituent of the causal relation. It now remains to address the *epistemological* question of why we are right to *hold* that causes typically precede their effects. This is worth our attention for two reasons. First, it is important to combat the tendency to conflate the questions. (One often comes across attempts to explain the direction of causation in terms of the nature of our concepts.) Second, the answer to the epistemological question is what distinguishes the view presented here from the conventional predetermination account.

We have just encountered one way of reaching the conclusion that our belief in the direction of causation in a posteriori. Regard the manipulability maxim, (M), as an analytic principle partly defining "causation", and combine it with the belief that rational choices tend to *precede* the events for the sake of which the choices are made. These principles, the second of which is a posteriori, entail that causation tends to be future oriented.

Let us consider a more popular route to the same conclusion. Suppose it often happened that light contracted in inwardly moving spherical wave fronts to points in space—the time reverse of what actually occurs. And imagine that apples formed out of the earth and leapt up into the branches of trees. How could these processes be explained? Could mere coincidence account for the remarkable coor-

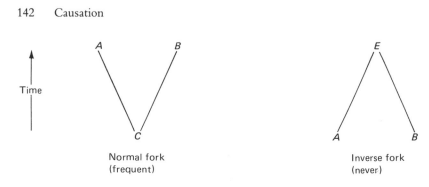

Figure 33

dination of separated wave fronts converging from infinity, or for the fact that apples are pushed from the ground by apparently random molecular vibrations in just the right directions and at just the right speeds to attach themselves to the conveniently receptive foliage? A tempting alternative would be to explain these things teleologically, by means of the hypothesis that the light waves are *destined* to be absorbed at a certain point and the apples *fated* to end their days in trees. Thus we might be a little inclined to postulate a form of backward causation.

This temptation comes from applying the familiar maxim, "Correlated events are causally connected," to a hypothetical pattern of events, known as an "inverse fork" (see chapter 4)—a pattern involving separated event types that are strongly correlated with one another and with a later event but not associated with any central, common antecedent. In the diagram on the right of figure 33, A and B represent parts of an inwardly moving wave front and E represents their ultimate absorption. In view of the symmetry of the situation, and the fact that the maxim of causal correlation is extremely well-entrenched, we are tempted to say that E causes A and B, rather than admit an uncaused correlation between A and B.

Thus one might invoke the maxim of causal correlation to explain our inclination to postulate backward causation. And in a similar way one might use it to explain why we believe in the predominance of forward causation. (Something like this idea has been endorsed by Reichenbach 1956, Dummett 1964, Mackie 1974, and Papineau 1985) Begin with the thesis that the direction of causation, in general, is fixed by the direction of causation in the special context of forks; then note that the maxim of causal correlation entails that in a fork the central event causes the separated correlated events; and, finally, bring in the vital contingent fact that there are many cases of so-called "normal forks" in which separated, simultaneous correlated events

are associated with an *earlier* central event but not with any characteristic *later* event, but the time reverse of this pattern—an inverse
fork—is never exemplified. In other words, the direction of causation
is analytically defined as the direction that would provide correlations
with causal explanation. This is the direction in which forks spread
out. And that, as a matter of contingent fact, happens to be the future.

Thus the collection of alternative 'substantive' accounts offers us a
range of different a posteriori answers to the question of why we
believe in the future orientation of causation:

1. "Causation" is defined—not in terms of time order—but, in
    part, by the principle that correlated events are causally con-
    nected, and this, given the fact that there are no inverse
    forks, determines that causation is future oriented.
2. "Causation" is defined through its association with our ex-
    perience of deliberation and control. More specifically, we
    define causation as that general relation between events that
    is exemplified when an event is deliberately brought about
    by free choice. But, because of the difficulty of identifying
    the antecedents of decision, our voluntary actions are per-
    formed only for the sake of future events. Hence the direc-
    tion of causation.
3. "Causation" is defined, in part, by the idea that a cause is in
    some sense 'ontologically more basic' than its effects. But,
    because of the knowledge asymmetry—very roughly speak-
    ing, the past is knowable and the future is not—we tend to
    think that the past has 'more reality' than the future (note
    our discussion of the tree model in chapter 2). And this leads
    to the idea that the past is causally prior to the future.

And perhaps there are more. But it is doubtful whether there is any
clear sense to the thesis that one, and only one, of these answers can
be correct. Here, I am endorsing Quine's idea that we ought to abandon the positivistic picture in which our beliefs about some phenomenon are divided into two disjoint groups, containing, on the
one hand, those principles that merely define our terms and are therefore irrefutable and, on the other hand, those that express substantive,
falsifiable, empirical claims. Instead, we should recognize that none
of our beliefs is entirely sacrosanct, although some play a more pervasive, permanent, and prominent theoretical role than others. As our
knowledge evolves, difficulties emerge in the current cluster of
beliefs; some have to be given up. But where the axe falls is not
constrained, as in the old picture, by some body of antecedent, once-
and-for-all stipulations. Rather, the decision about how to proceed is

shaped by the relative attractiveness of the various global theoretical positions that each option would result in. Applying this idea, we must renounce the view that some one of the preceding principles is part of the 'analytic definition' of "causation". Rather, we can suppose that the concept of 'causation' involves all of the principles on an equal epistemological footing.

Most scientific concepts are like this. Consider, for example, the classical idea of 'straight line'. One of its application criteria was 'satisfying the axioms of Euclidian geometry'; another, 'being the possible path of a light ray'; another, 'being the possible path of a freely moving particle'; and another, 'being the locus of a stretched string'. Within classical physics these are extensionally equivalent— all supposedly applying to the same entities. But with the collapse of that theory, the question arises as to which of these nonequivalent criteria pick out straight lines. In fact, what happened was that the geometrical criterion was abandoned and the others were preserved. Note, however, that this decision was motivated by considerations of theoretical economy within the new physics. There was nothing in our earlier, classical use of the straight line concept that showed that it would be right to make that decision. Therefore, in explaining why we believed that light travels in straight lines, it would have been wrong to cite convention and also wrong to pin responsibility on just one of the principles from which this belief may be derived. The right answer is to show how the belief is embedded in a large cluster of mutually supporting commitments.

I am suggesting that our attitude toward the direction of causation is similarly explained by its integrated position within a cluster of beliefs, including:

    i.   Causes typically precede their effects.
    ii.   Correlated events are causally connected.
    iii.   Choices are made for the sake of what they might cause.
    iv.   The causes of a present event are knowable, but its effects are not.

It is a contingent fact, partly a consequence of the scarcity of inverse forks and partly stemming from our experience of deliberation, that these elements are roughly consistent with superficial observation and seem able to coexist with one another. But, strictly speaking, they are not perfectly harmonious. We do have real knowledge of the future, we acknowledge that scientific advance could turn up occasional cases in which correlated events are not connected by prior causes, and, we come to see that in Newcomb-like decision contexts acts may reasonably be performed for the sake of earlier events. In response to such facts the cluster of principles requires modification. Overall theoretical economy is promoted by recognizing that ii, iii, and iv are merely superficial approximations to the truth and by incorporating i within a characterization of causation along the lines of our predetermination-chain theory. However, although relegated to the category of "approximate truths", the maxims of correlation, deliberation, and knowledge remain part of our concept and play a significant role in bolstering the belief that causation is future oriented.

9
Explanation

Explanations of phenomena, descriptions of how or why things occur, usually refer to conditions that *precede* the events to be explained. Cases where this is not so, where the explanation is given in terms of subsequent conditions, are certainly not unheard of but tend to provoke discomfort. They either invite reformulation—"He jumped to avoid the bull" becomes "He jumped because he *wanted* to avoid the bull". Or they are simply unacceptable—"He became president to fulfill his destiny". Thus explanation appears to be time-asymmetric, and our problem is to say why this should be so. To this end, let us try to achieve a general understanding of the nature of explanation. Remember that the aim is not to give definitive analyses of 'causation', 'explanation', 'law of nature', and 'counterfactual dependence' but rather to uncover just enough of their structure and interaction to resolve the problems of time asymmetry.

According to Hempel's (1948) classic account, to explain a fact is to show how, under the circumstances, it was only to be expected. And this is accomplished by deducing the fact from various premises about the prevailing conditions and laws of nature. Consequently Hempel's theory is known as the Deductive-Nomological (D-N) model.

Many cases of explanation do indeed conform to this analysis. For example, we might explain why a certain substance is burning with a green flame by pointing out that it is a copper salt and citing the law that all copper salts burn with a green flame. Nevertheless, it is well-known that the D-N model is wrong. On the one hand, there are perfectly decent explanations that are not covered. And, on the other hand, there are clear instances of the model that do not qualify as explanations. Thus conformity with the D-N ideal is neither necessary nor sufficient for an explanation. Let me quickly review some of the familiar types of counterexample (van Fraassen 1980).

First, we have statistical explanation: for example, Q. "Why did

Smith contract leukemia?" A. "Because he works with plutonium." This could be the explanation even if there is only a low probability of leukemia associated with the particular dose of radiation to which Smith was exposed. Therefore explanations need not be deductive proofs.

Second, it is not unusual to explain something by reference to just a single fact (as in the previous example). Now this point on its own does not refute the D-N theory, since one might suppose that the rest of the deductive (or statistical) argument is tacit—perhaps too obvious to mention. But that defense is inadequate because it would often be unavailable. For example, one might attempt to explain the demise of British feudalism in terms of the weakness of the monarchy, without claiming to be actually in possession of a system of facts and laws (even probabilistic) from which the event could have been inferred. Therefore even nonstatistical explanations may fail to conform to the D-N ideal.

Third, there is the notorious asymmetry problem, showing that valid instances of the D-N model need not be explanations. For instance, we may explain the length of a shadow by deducing it from the height of a flagpole, the angle of the sun, and the law that light goes in straight lines. However, it would be equally possible to deduce the height of the pole from the length of its shadow and the other information. Now, as Sylvain Bromberger (1966) observed, this is a good way of *finding out* the pole's height, but it surely would not tell us *why* it is that high. Thus a deductive argument may give us reason to believe a fact without thereby explaining it.

Fourth is a problem of relevance: for example, "Smith recovered from his cold because he took vitamin C, and almost all colds clear up within a week of taking vitamin C" (Salmon 1971). Here, a fact is deduced from a feature of the circumstances and a lawlike generalization, just as prescribed by Hempel's theory. But suppose that we believe this generalization only because it follows from "All colds clear up within a week." In that case the purported explanation is unacceptable. We regard the fact cited as irrelevant, even though it can be made to figure in a D-N account of what is to be explained.

Finally, a D-N answer can be singularly unhelpful: for example, Q. "Why was Al in the bank vault at 3 am?" A. "Because he was there at trillionth of a second before 3 am, and nothing travels faster than light" (Putnam 1978). Even though this conforms to the D-N model, it would not satisfy anyone who was demanding an explanation.

In the face of these difficulties it is clear that Hempel's account must be either scrapped or revised. But the problems also have a constructive aspect. In revealing the basic flaws in the D-N conception of explanation, they point us in the direction of a better theory.

One way of conceiving of what is wrong is this. Evidently the usual way of explaining an event consists in specifying its causes (Salmon 1984, Lewis 1986). This almost goes without saying. But it means that the D-N model of explanation can be no more adequate than the very crude nomological determination theory of causation that is embodied in it. And we saw, in chapter 8, some of the difficulties that such a theory is subject to. For example, it was obvious that a fact may enter (together with laws of nature) into the deduction of another fact without being a cause of it. Thus the trouble with Hempel's account is its implicit committment to a defective theory of causation.

In this light, the counterexamples to Hempel's theory are not at all surprising. If Smith's radiation overdose caused the leukemia, even if there was only a small chance that this would happen, then it may be cited in the explanation of his illness. Similarly, if I believe that a weak monarchy caused the demise of feudalism in Britain, then I don't need to claim knowledge of 'historical laws'—or of any laws at all—to explain that event. We can believe that something causes a fact without being currently in possession of a system of laws in which the mechanism of the causal connection may be exhibited. Third, the asymmetry problem arises because nomological determination, by itself, does not imply causation. And, fourth, insofar as we don't think that the vitamin C was a cause of recovery, we won't cite vitamin C when trying to explain the cure. The overall moral here is that in order to deal with these four counterexamples, Hempel's approach must be modified: first, by making explicit that the most prominent form of event explanation is a specification of causes and, second, by employing the refined predetermination-chain account of causation, advocated in chapter 8, instead of the crude analysis implicit in the D-N model.

This goes not just for particular facts but also for many general laws. In order to explain why all A's are B, we must say how it comes to be the case that this is so—and this is often a matter of specifying its causal antecedents. For example, the laws of thermodynamics may be explained in terms of statistical mechanics by virtue of various causal theses: for example, that the pressure of a gas is brought about by molecules banging into the container walls, and that temperature uniformity is produced by intermolecular collisions, which tend to boost the speeds of slowly moving molecules and to slow down the fast ones. Typically, generalizations about a certain kind of entity (stars, human beings, samples of water, neutrons, etc.) are explained by theories that describe the constituent structure of the entities in question and show how the features to be explained result from the interaction of those constituents. In other words, to explain the truth of the generalization, "All *A* are *B*," it would suffice to say what conditions must obtain in order that an entity of type *A* be constituted, and then to show that any instance of such conditions will also cause the presence of property *B*.

2. Pragmatics

Our fifth counterexample has a different source, and throws into relief a *pragmatic* dimension of the concept of explanation. In justifying our response in that case, we are inclined to protest that *Al*'s location at a trillionth of a second before 3 am was already obvious (given the question at issue) and is no less puzzling (and for exactly the same reasons) than his location at 3 am. So you can't properly explain one in terms of the other. This reaction brings out the fact that explanation is a social act: a transfer of information from one person to another. Therefore, not surprisingly, the value of the act—whether it counts as a good or a bad explanation—will depend to some extent on the antecedent beliefs and interests of the participants. To explain a phenomenon, one describes how it was produced. However, the selection, from among all the causal antecedents, of those to be explicitly cited, is determined by what the recipient of the explanation wishes to know about the causation of the phenomenon. This contextual variable may be decomposed into a number of ingredients, including:

> 1. *Contrast class.* In asking why something happened, one wants to know why *it* happened, *rather than certain alternatives* (Dretske 1972). For example, why did Smith lend Jones $5, rather than give it to him? Or, why did Smith, rather than Brown, or Black, lend Jones the $5? Or, why $5, rather than $6, or $7? Clearly, what will qualify as a good explanation in this case depends on which particular contrast class the questioner has in mind.
> 2. *Background assumptions.* A good explanation must provide *new* information. Typically the questioner will already know something about the causal antecedents of the event to be explained, and obviously does not want to be given merely what he already has.
> 3. *Surprise removal.* It may be that the event to be explained is especially surprising given some of the background assumptions.

In that case the explanation must address this surprise, by either citing further factors, of which the questioner was previously unaware, that eliminate or diminish it or denying the assumptions from which the surprise was derived.

4. *Interest relativity.*   Often the questioner's concern is confined to a special subset of an event's causes. This contextual variable is illustrated nicely in a passage by R. G. Collingwood (1940, p. 299):

> For example, a car skids while cornering at a certain point, strikes the curb, and turns turtle. From the car-driver's point of view the cause of the accident was cornering too fast, and the lesson is that one must drive more carefully. From the county surveyor's point of view the cause was a defect in the surface or camber of the road, and the lesson is that greater care must be taken to make roads skid-proof. From the motor manufacturer's point of view of the cause was defective design in the car, and the lesson is that one must place the centre of gravity lower.

When one or more of these pragmatic desiderata is not satisfied, then the explanation is defective. One might be tempted to worry about precisely how such imperfect explanations should be characterized: as *false*, merely *inappropriate*, or perhaps as *unsuccessful attempts at explanation*. But I suspect there may be no definite or useful answer to this question, and I won't discuss it.

The contextual constraints on decent explanation have been rightly emphasized by Bas van Fraassen (1980). However, although they should certainly not be neglected, it is possible to overstate their significance, and it seems to me that van Fraassen does just that when he suggests that a proper appreciation of pragmatics will solve the asymmetry problem.

Consider again the flagpole example. According to Hempel's model something explains the truth of a statement if it is a deduction of that statement from laws of nature and particular circumstances. Thus we can explain the length of a shadow by citing the height of the pole that casts it, the inclination of the sun, and the law that light travels in straight lines. However, we could equally well deduce the height of the pole from the length of the shadow, the sun's inclination, and the light law. Yet, contrary to Hempel's theory, this would surely not qualify as an explanation. Van Fraassen's idea is that the problem may be solved by attention to the pragmatics of explanation—to the way in which the interests of the speaker and hearer determine which features of the causal environment of what is to be explained it is appropriate to mention. Thus in Collingwood's example, what is selected as *the* cause of the accident is, variously, negligence, road surface, and car design, depending on the context, though of course objectively all of these factors, and many more besides, are causally efficacious. Similarly, says van Fraassen, there is no objective asymmetry in the flagpole case. For in some contexts the length of the shadow would indeed help to explain the height of the pole. For example, one might be told that the pole was deliberately designed with a particular height so that it would cast a shadow of some desired length at a particular time of day.

However, the underlying asymmetry has not really been eliminated. Granted, there is a possible explanation of the height of the pole that does cite the length of its shadow—but only as a way of indirectly designating the *causes* of the pole's height, namely, the *desire* of the manufacturer to produce that shadow length and his *beliefs* about light, solar inclination, and trigonometry. The shadow length itself is not directly relevant since it is an *effect* of what is to be explained; but it is indirectly relevant, and is therefore mentioned, because it serves, in the context, to identify some of the directly relevant (i.e., causal) factors. Note that it is only in the very special circumstances of deliberate planning that we can explain a pole's height in terms of its shadow length, whereas it is almost always legitimate to explain the shadow length in terms of the pole's height. Thus the real key to the asymmetry of explanation is the asymmetry of the causal relation. And that asymmetry is an objective fact, not a matter of pragmatics. Its origin, according to the theory sketched in chapter 8, is time order. The *earlier than* relation appears as a constituent of the causal relation, and this is what accounts for both the prevalent direction of causation and the asymmetry of explanation.

3. Functional explanation

Although pragmatic considerations do not solve the asymmetry problem, they do help in the treatment of another longstanding controversy which is closely related to the asymmetry question. I have in mind the issue of functional explanations. And I am concerned specifically with the questions: "Are they legitimate?" and "Do they display an abnormal temporal orientation?"

Consider, for example:

Q. Why do humans have hearts?
A. To pump blood.
Q. Why do societies have ethical norms?

A. To promote harmony and cooperation.

Q. Why do humans dream?

A. To permit sleep.

The aim in such cases is to account for a property, P, of systems of a specified type, S, by indicating some causal consequence, F, of P. F is the so-called *function* of P's presence in S. So, in our first example, $P =$ having a heart, $S =$ human beings, and $F =$ the circulation of blood. Explanations of this sort are frequently given; yet they have encountered a great deal of philosophical criticism. The two main objections have been as follows:

> 1. *The explanatory irrelevance of effects.*   An explanation of a phenomenon is an account of how it was brought about—a specification of at least some of its causal antecedents. But in the preceding cases nothing is said about what *causes* the presence of P in S. Rather, all we are told is that one of P's *effects* is F. Thus, that statement does not explain the presence of P. For example, the fact that a person's heart pumps his blood does not explain how he came to have one.
> 2. *The problem of functional equivalents.*   Something, $P^\star$, other than P, could also cause F; therefore the presence of P, rather than $P^\star$, cannot be accounted for simply by reference to that effect. For example, a mechanical device would pump blood; so the fact that hearts do this cannot suffice to explain their presence.

However, these objections fail to undermine the propriety of functional explanation. They misidentify the intended explanatory thesis and neglect the role of contextually supplied background information. Let me substantiate these claims.

A functional explanation attempts to say why it is that all (or most) systems of type S have property P (or sometimes why some particular S has P). The explanatory thesis is a causal claim. It says that the thing to be explained is caused, in part, by the following general fact: that the presence of P in an S tends to produce, or facilitate, feature F. How such a causal claim may be established will of course vary, depending on the subject matter. One possible way of doing it would involve showing that the presence of feature F is essential for the survival of any S. Thus we might explain why societies have ethical norms by arguing that in their absence the level of cooperation required for stability would not be attained. Another general strategy would be to show that F fosters the reproduction of S's, and that P is an inherited characteristic. The important point, however, is that a functional explanation, once spelled out. is clearly nothing but a kind of causal explanation. The fact we are citing (roughly, that P produces F) is not an effect of what we are trying to explain. Therefore the complaint that effects are explanatorily irrelevant, though true, is beside the point.

The second objection was that 'All S's are P' could not be explained by 'P produces F', because there are other properties, say $P^\star$, that also produce F, and yet it is not the case that all S's are $P^\star$. But this is like arguing that striking a match can never be a cause of its lighting since struck matches don't always light. What is wrong of course is that the importance of having the right surrounding circumstances has been forgotten. It is in virtue of further, unspecified conditions that P's tendency to produce F is able to play its alleged causal role. For example, it may be a vital aspect of these conditions, in certain cases, that at least some systems of type S with property P are naturally created, whereas systems with $P^\star$ are not. Thus it is perfectly possible for there to be conditions that enable P's relationship with F—but not $P^\star$'s similar relationship—to have certain consequences. Sometimes it may be important to mention explicitly those further conditions. However, on other occasions our explanation can be quite in order without mentioning these facts. There may be shared background knowledge of them, in which case it is good enough simply to point out that P produces F.

Thus functional explanations are ordinary causal explanations. They invoke no strange forms of past-oriented, teleological influence. Therefore their existence does not diminish the extent to which explanation is asymmetric with respect to time.

4. Does explanation depend on causation, or vice versa?

I have been arguing that to explain an event is to point out its significant causes—where significance is determined by a range of pragmatic considerations. Given this thesis, it might appear that explanation should be *defined* in terms of causation, and that the concept of causation is more fundamental than the concept of explanation. For the following reasons, however, I think that this way of thinking is incorrect.

In the first place, not all explanations are explanations of why or how physical events occur. Some concern facts in philosophy, linguistics, mathematics, and other domains, involving phenomena that do not enter into causal relations. Obviously explanations of such phenomena must be noncausal. Second, a conception of explanation that is broad enough to encompass these fields might well be capable of providing a basis for our notion of causation. For, if we were armed with a general, noncausal account of explanation, we could then try to refine the characterization of a *cause* as an *explainer of events*—thus employing our initial thesis to define causation in terms of explanation rather than the other way round. And third, the advisability of this particular direction of theoretical priority is reinforced if we look back to the analysis of causation presented in chapter 8. For that account relied on the notion of *basic* law (to characterize 'chains of direct determination') and therefore presupposed the concept of explanation. Let me explore these points.

Explanation often takes place in noncausal domains. For example, the principle of utility might be proposed as an explanation of why it is wrong to hurt people; the syntactic constraint, "reflexives must be bound within their own clauses," tells us why "John wants Mary to like himself" is ungrammatical; the laws of probability explain why a large random sample is usually representative of its population. Even in physics, where causal explanations predominate, there are noncausal explanations of certain general laws. For example, one might cite the structure of spacetime to explain why time travel is possible, or use the principle of conservation of energy to account for the properties of levers. Thus we certainly have a conception of explanation that is noncausal.

In each of these cases the fact explained is said to follow from facts that are held to be, in some sense, 'more basic'. We seem to have in mind something like the following picture (Friedman 1974). Given any domain of inquiry, we expect there to be a relatively small set of simple principles that logically entail all other facts in the domain. These are regarded as the most basic facts. Then other facts can be assessed in terms of their 'distance' from this core. Very crudely, we say that P is more basic than Q if the simplest deduction of Q from the core of most basic facts 'passes' through P. Thus a certain asymmetry is suggested—namely, if P explains Q, then Q does not explain P.

So much for an extremely rough picture of the 'logic' of explanation. Now, our problem is to understand why, in the domain of events, explanation becomes a specification of causes. In this domain, the core of most basic facts will contain a set of simple generalizations—the basic laws of nature (see chapter 10). But these will not suffice for the deduction of all the physical facts. It will be necessary to include within the basic core a set of *particular* facts, and we tend to cast in this role the initial conditions of the universe. (For the sake of simplicity I assume that there is such a state and that determinism is true.) But why do we explain by reference to *initial* conditions? Why do we regard *earlier* conditions as more basic than later ones. At least part of the answer lies, I think, in the fact that earlier sets of circumstances are simpler (in a certain respect) than later ones. This is due to the familiar fork asymmetry—correlated, simultaneous events are linked to a characteristic prior event that determines them both, but there need be no characteristic subsequent event. Consequently correlations can be derived from a more unified earlier condition. So there is a gain in the simplicity of our characterization of the world as we move back in time from states of correlation to their unified determinants. In other words, the earlier, central event in a fork allows a unified derivation of the separated correlated events and is therefore explanatorily more basic than them. However, this is not the whole story. A complete account of the direction of explanation would have to mention a *cluster* of factors, paralleling those maxims invoked earlier to explain our faith in the direction of causation. Specifically, we might say that we simply believe that *earlier* events are more basic than *later* ones, that the known seems 'more real' and more basic than the unknown, and that our decisions are more basic than the events that are nomologically connected to them. It's not clear that we can deny any of these maxims a role in fixing the direction of explanation.

In chapter 8 I sketched a construction of the causal relation that began with the idea of 'chains of direct nomological determination', and then added constraints of continuity and typical time order to provide the chains with a definite causal interpretation. I argued that each elementary link in these chains is an instance of an explanatorily basic law of nature. And now we have just seen that the time-order constraint may also be derived from a view of what is explanatorily basic. These considerations suggest that we can characterize the nature of causation from an independently definable notion of explanation, but not vice versa. Thus there is reason to suppose that 'explanation' is theoretically prior to 'causation'.

10
Counterfactuals

---

1. Goodman's problem

Counterfactuals concern what would happen, or have happened, in specified hypothetical circumstances. For example, one might say of an unused match, "if struck, it would have lit". The analysis of such statements is important. For, in the first place, they are pervasive: the meanings of countless scientific predicates (soluable, malleable, ductile, explosive, etc.) appear to involve commitments to what would occur under certain possible circumstances. And second, the notion of counterfactual dependence is bound up with other puzzling concepts, such as 'law', 'cause', and 'explanation'. Yet, despite considerable attention by philosophers of science in the last forty years, there has emerged no satisfactory way of describing what must be the case for such conditional statements to hold. In particular, it is evident that the truth of "If p were true, then q would be true" is not determined solely by the truth values of p and of q. Typically p and q are both believed to be false, yet it remains open whether or not the counterfactual claim is acceptable.

One of the first sustained attempts to grapple with this matter is Nelson Goodman's classic essay, "The Problem of Counterfactual Conditionals" (1946). He suggests, there, that a counterfactual, $p \;\square\!\!\rightarrow\; q$ is true when its antecedent, p, nomologically requires, given prevailing conditions, the truth of its consequent, q. Or, in other words, that the antecedent, p, conjoined with some set of facts, S, and laws of nature, L, deductively entails the consequent, q:

$$p \;\square\!\!\rightarrow\; q \quad \text{if and only if} \quad (p \,\&\, S \,\&\, L) \rightarrow q$$

However, this idea has foundered on two major obstacles: the so-called "cotenability problem", and the explication of "laws of nature". I shall consider the first of these difficulties immediately, and postpone the second one until section 4.

It is evident that some restriction must be placed on the contents of S. For suppose the antecedent is in fact false, as in the case of the unstruck match. And suppose that not-p—which, in that case, is a fact—were to be included among the prevailing conditions, S. Then ($p \& S \& L$) would be a contradiction, entailing any q whatsoever. In this way all counterfactuals (with false antecedents) would be made trivially true.

In the attempt to overcome this sort of difficulty, it turns out that membership in S has to be restricted to facts that would not be altered by the truth of p. That is to say, S must be required to satisfy the condition: $p \: \square\!\!\rightarrow S$. Thus Goodman is forced to acknowledge that the restriction clause must itself involve counterfactual conditionals, thereby rendering the whole analysis circular. For example, we accept

   If the match had been struck, it would have lit

because the match was dry and oxygen was present. We do not, however, accept

   If the match had been struck, there would have been no oxygen

even though the match was dry and did not in fact light. Goodman found, as I have just said, that he could match these intuitive judgments only by requiring that the background facts, S, be *cotenable* with the antecedent (i.e., be such that they would remain true even if the antecedent were true). Given such a rule, the non-lighting of the match cannot be invoked as one of the background facts. But, as Goodman points out, this solution is no good at all, since it relies upon the very notion we were trying to explain.

2. A causal/explanatory theory of counterfactuals

Can we use the notion of causation to solve this problem? Causation and counterfactual dependence are intimately related ideas, and it is natural to want to explain one in terms of the other. This has been tried by Lewis (1973b), who analyzes causation in terms of counterfactual dependence. However, I believe that his approach puts things the wrong way round and that this basic error yields a host of counterintuitive consequences. Happily, there is a decent alternative. Our predetermination-chain theory of causation, based on the concept of law, is at hand (see chapter 8). If this account is roughly right, then there is no need for a prior grasp of counterfactuals. On the contrary, the way is clear to use causal concepts in specifying the meaning of counterfactual conditionals. Let me now indicate how this can be done. In the final section of this chapter, I shall give extra support to this approach by describing and criticizing Lewis's theory in some detail.

My plan is to use causation in a solution to the cotenability problem, thereby enabling a revival of something like Goodman's strategy. He proposed that the counterfactual $p \;\Box\!\!\rightarrow\; q$ is true just in case the truth of p would, *given the circumstances S*, nomologically determine the truth of q. But, as we have seen, this theory runs into trouble over the word "circumstances". For it looks as though the only adequate account of that notion would itself employ counterfactuals, making the analysis viciously circular.

I want to suggest, however, that the relevant conception of 'circumstances' may be described in terms of causation, rather than counterfactual dependence. Specifically, we should consider the circumstances S, (in which the truth of the antecedent is to be supposed) to consist of any facts that are *not* causes of, caused by, or nomologically determined by $-p$, the falsity of the antecedent. Thus, in order to validate $p \;\Box\!\!\rightarrow\; q$, imagine a change in the world (p instead of $-p$), hold fixed things that are logically, nomologically, and causally independent of the changed fact ($-p$), and observe that the consequent (q) would be determined. In other words, consider a possible world containing all phenomena that are neither causes nor effects of (nor nomologically determined by) the events described by $-p$ but containing p. If (and only if) our laws of nature determine that q obtains in that world, then the counterfactual, "If p were the case, then q would be the case", is definitely true. If our laws entail merely a certain probability, x, that q obtains, then the probability is x that the counterfactual is true. I think that this resolves the cotenability problem—and without a hint of circularity since counterfactuals are not themselves relied on in our account of causation.

Consider, for example, the case of the unused match. How can we accommodate the fact that we accept

If the match had been struck, it would have lit

but not

If the match had been struck, there would have been no oxygen

The answer, in terms of our causal theory, is obvious. The actual nonstriking of the match caused neither the presence of oxygen nor the dryness of the match. And it was not caused *by* either of them. Therefore both facts qualify to be among the circumstances of the supposed striking, and so the lighting will indeed be determined. On the other hand, the nonstriking of the match *was* a cause of its not lighting. Therefore the nonlighting fails our condition for inclusion among the circumstances of the supposed striking. Thus, although the nonlighting would, given the antecedent, imply an absence of oxygen, this fact does not validate the second conditional.

In the present account, no hypothetical *particular* fact can counterfactually imply the violation of a law of nature. This is because in determining what depends counterfactually on an event *E*, the theory tells us to hold fixed anything that is logically, nomologically, or causally independent of *E*, and any law of nature will meet that condition. Thus there are no true counterfactuals of the form. "If *E* had not occurred, then *L* would be false", where *E* is a particular event and *L* is a law of nature. However, the theory does not exclude *counterlegals*, which are counterfactuals whose antecedents hypothesize the falsity of laws. In such a case we must hold fixed any event whose occurrence was not governed by the laws whose violation is supposed, as well as any further laws that neither explain nor are explained by them.

Notice that my characterization of the meaning of counterfactuals is given in terms of 'assertibility conditions' rather than truth conditions. In other words, instead of providing a traditional style of analysis for "$p \;\Box\!\!\rightarrow q$", I have specified the circumstances in which one ought to believe to degree x that $p \;\Box\!\!\rightarrow q$. That is to say, I have described the conditions for

$$\text{Prob}\,(p \;\Box\!\!\rightarrow q) = x$$

My reason for taking this approach is an inclination to agree with Robert Stalnaker (1984) that $p \;\Box\!\!\rightarrow q$ might be true even if p, given the prevailing circumstances, would certainly not determine q. I would say, for example, that there is a probability of one half that if a certain fair coin had been tossed it would have landed heads up. But in these cases (when they are genuinely random) there appear to be no facts, more basic than the counterfactuals themselves, in virtue of which they are true. Let me emphasize, however, that even if I am wrong on this matter, the main point remains unaffected: namely, that the problem of specifying 'prevailing conditions' need not introduce a circularity into the analysis of counterfactuals, since it may be solved by employing the concept of causation.

Notice also that my proposal is specifically tailored to counterfactuals that concern events. In order to obtain a more general account, we must generalize the cotenability condition. Instead of supposing that the 'prevailing conditions' be facts that are not *causally* explained by, or *causal* explanations of, the falsity of the antecedent; we must drop the restriction to *causal* relations and say, more generally, that the 'background facts' are facts that are not explanations of, or explained by $-p$—in whatever sense of explanation is appropriate to the domain in question. Thus, it would be more accurate to call our account "an explanation theory of counterfactuals". For causation enters the picture only because, in the domain we are mainly interested in, explanation is a specification of causes.

3. Is counterfactual dependence time-asymmetric?

One big difference between the preceding account and conventional wisdom about counterfactuals is over the alleged *direction* of counterfactual dependence. It is often said that true counterfactuals are typically future-oriented, concerned with what would happen *subsequently* if some hypothetical event were to occur. Indeed, this alleged feature of counterfactual dependence is essential to theories, such as Lewis's, that analyze causation in terms of counterfactual dependence and that attempt to explain the direction of causation in terms of the temporal properties of counterfactual dependence.

Proponents of this view will generally concede that there are true past-oriented counterfactuals like

> If the match had lit, it would have been struck

But they claim that in such cases a *nonstandard* notion of counterfactual dependence is employed (deriving, in Lewis's theory, from a nonstandard metric of possible world similarity). On our account, however, the counterfactual "if. . . then. . . " is univocal, and a fact may depend on a later event just as easily, and in the same sense, as it depends on an earlier event. Thus on the present account counterfactual dependence is not asymmetrical with respect to time.

This thesis might seem wrong. For one must admit that it is unusual to find a statement of the form

> If p were true, then q would have been true

when q is about events earlier than the time of p. However, the difficulty here disappears as soon as we recognize that past-oriented counterfactuals are only rarely, and with considerable strain, expressed in precisely that form. Instead, we tend to say

> If p were true, then q would *have to* have been true

or more often,

> p would be true, only if q had been true

Thus future- and past-oriented counterfactuals tend to be formulated in different ways. Nevertheless, the underlying relation of counterfactual dependence is time-asymmetric.

In denying that there is any need to proliferate senses of the counterfactual conditional, I follow the lead of Jonathan Bennett (1984).

He points out that the temptation to postulate an ambiguity comes from the existence of pairs of seemingly acceptable conditionals whose apparent incompatibility may be dissolved by acknowledging an equivocation. For example, there are circumstances in which both

1. If he had jumped off the building, he would have been hurt

and

2. He would have jumped, only if there had been a safety net

would seem to be true. And a way to reconcile them is to suppose that different kinds of dependence are involved.

However, as Bennett rightly says, the intuition that both statements are true is not strong enough to justify a substantial increase in theoretical complexity. Rather than introduce multiple senses of the conditional, we can simply insist that in fact, depending on the circumstances, only one of those claims is true. For example, if the man were on the verge of jumping and was barely talked out of it, then statement 1 looks plausible. But if he merely went up on the roof to admire the view and was jokingly challenged to jump, then 2 seems like the right thing to say. Moreover we can soften the conflict with our initial intuition even further, by recognizing that the *full* antecedent of a counterfactual conditional may not always be made explicit. Thus someone might say 1 when what he means is, "If he had jumped and there had still been no safety net, then he would have been hurt"—thereby reconciling the thought behind 1 with 2.

This kind of strategy is taken to extreme lengths in a theory of counterfactuals propounded by van Fraassen (1980). According to his view, a counterfactual conditional is simply a logical entailment claim whose antecedent is only partially explicit and residually tacit. That is,

$$p \,\square\!\!\rightarrow q \quad \text{if and only if} \quad (p \,\&\, F) \rightarrow q$$

where F is a body of facts whose content is decided on by the speaker but not expressed. Thus in different contexts, depending on what the speaker has in mind, a single counterfactual may have different truth values. Moreover this context dependence is taken to imply that counterfactual statements are not scientifically objective.

In response to this idea, I would suggest that van Fraassen overestimates the extent to which a speaker is required to dictate the circumstances in which his antecedent is to be supposed. The speaker does not need to intend that the circumstances include facts that are causally independent of $-p$, for this is already implicit in the meaning of the conditional. Moreover, if our linguistic practice were as van

Fraassen says it is, then we would be hard pressed to explain disagreement over counterfactual theses. Surely these conflicts of opinion are not all simply a result of failure to appreciate what the speaker is assuming. Moreover, the disagreements are typically settled by reference to empirical evidence; and again this would be inappropriate if counterfactuals were compressed assertions of logical entailment.

This is not to say that the speaker's intentions have *no* role to play. As we have seen, they can determine which particular counterfactual statement is expressed by the utterence. But radical conclusions do not follow from this concession. First, it does not transform counterfactuals into entailment claims. And second, it renders counterfactuals no less scientifically objective than other theses that are not always spelled out explicitly.

4. *Laws of nature*

Besides the question of cotenability a further difficulty with any Goodmanesque view of counterfactuals, such as the one I am proposing here, is the need to explicate the notion of *law*. For a counterfactual definitely holds only when its consequent *must*, by law, obtain, given the circumstances. But what are we saying when we maintain that some statement does not just *happen* to be true, but is a law of nature? What distinguishes laws, such as

All emeralds are green

from non-laws, such as

All the coins in my pocket are dimes

We must confront this question, not only for the sake of understanding counterfactuals, but also in order to ground the analysis of causation in terms of law, that was given in chapter 8.

To begin with, Goodman points out that no purely syntactic criterion will distinguish laws from other facts. Both of the preceding statements, for example, are universal generalizations. Moreover any particular fact, "*k* is an *F*", is logically equivalent to the syntactically general, "Everything identical to *k* is an *F*". Instead, Goodman proposes an epistemological answer derived from Hume. What is special about laws, he says, is their role in enabling us to make predictions.

This idea reflects the analogy between projecting a generalization onto purely *hypothetical* entities, and projecting it onto *unobserved, actual* entities. Suppose that statements that serve one purpose also serve the other. Then the condition for being lawlike (i.e., a law if true), and thus capable of sustaining counterfactuals, will be a suitability for making predictions. In particular, a generalization of the form "All A's are B" will count as a law, if and only if it is projectible—that is, the observation of A's that are B provides reason to believe that unobserved A's are B.

But when is this so? The problem of distinguishing projectible and nonprojectible generalizations—called by Goodman, "the new riddle of induction"—is too large and difficult a topic to discuss here. However, a plausible preliminary view of the matter (Goodman 1955; Horwich 1982) is that hypotheses are projectible when they employ terms that are *natural* (that is to say, entrenched words such as "green" rather than 'defined' words such as "grue"), and when they are nevertheless syntactically *simple* (like "All A's are B", and unlike "All A's are either sampled and B or unsampled and $-B$"). Given some such account of projectibility, Goodman's account of laws reduces to the idea, roughly speaking, that laws are naturally simple generalizations.

I think we should accept the Humean idea that there is an intimate relationship between lawlikeness and being a naturally simple, projectible generalization. But it oversimplifies matters simply to identify these notions. To see this, notice that non-laws are often projected. For example, suppose that a box of matches is dropped in a puddle, and there is then some question as to whether they have been ruined. I try a few from various parts of the box, and all of them light quite easily. On this basis I confidently conclude:

Every match in the box is dry

which is clearly not a law of nature but nevertheless projected from a small sample. Similarly consider Hempel's (1966) notoriously troublesome example:

All objects made of pure gold have a mass of less than 100,000 kg

This isn't a law. But its high credibility does not depend on the observation of every single gold object.

Thus there are true, projectible generalizations that are not laws of nature. And this should not be surprising. All it takes for a generalization to be projectible is the belief that its truth would not just be a coincidence, but that there would be some uniform reason for conformity. In other words, a projectible generalization cannot be wholly accidental. But there are many generalizations that are true because of some combination of laws and particular fact. We might call them "semi-laws". In such cases the nomological component may render them projectible; yet the component of particular fact will disqualify them as pure laws of nature.

In response to this objection it might be said that the alleged counterexamples are not *absolutely* projectible generalizations (confirmable *solely* by the discovery of positive instances) but rather merely *relatively* projectible (confirmable by positive instances, only given the right background of further assumptions). For example, the discovery of dry matches confirms the hypothesis that all the matches are dry only relative to a context in which it is known that the box was dry initially, then dropped into a puddle, and so on. So "All the matches are dry" is merely relatively projectible. Therefore one might argue that lawlikeness should be identified with absolute projectibility, so the alleged counterexamples will be disqualified. However, this refinement of Goodman's idea cannot be right. For, whether or not a hypothesis H is absolutely projectible is an a priori matter, whereas whether or not H is lawlike is a posteriori. Let me elaborate. Most evidence claims, in ordinary language, are dependent for their truth or falsity on the theoretical context in which they are made. Thus a particular claim may reasonably be asserted at one time and then denied later on, given a background of new theoretical beliefs. However, the possibility of such variation is eliminated in the case of evidence claims that explicitly specify their theoretical background. In other words, whereas "E confirms H" may be true when asserted at one time and false when asserted later, because of a change of theoretical background from $B1$ to $B2$, the statements "E confirms H relative to $B1$" and "E does not confirm H relative to $B2$" are not subject to such variation. In particular, if we let $B0$ represent the 'null' background, in which no further facts are presupposed, then "E confirms H relative to $B0$" is not subject to revision in light of new discoveries. Thus absolute projectibility is an a priori matter. But lawlikeness is not. It is perfectly reasonable to believe at one time that H is lawlike (i.e., that if it is true, then it is a law) but subsequently in the light of new information to reach the conclusion that even though H is true, it is *not* a law. For we might come to think that H's truth would be in part a consequence of some quite accidental particular fact. Thus whether H is lawlike depends to some extent on which other theories are true. Therefore we cannot identify lawlikeness with absolute projectibility.

This discussion indicates that besides the notion of '*naturally simple (projectible) generalization*', a further ingredient of the concept of law is *explanation*. More specifically, it seems that in order to arrive at the set of laws, we must somehow restrict the class of naturally simple generalizations; and a plausible way to do so is by means of the following 'explanation requirement':

>Laws are explainable, if at all, only in terms of other laws—and never in terms of facts that arn't laws.

Assuming an understanding of explanation (from chapter 9), we can use this principle to whittle down the set of true, simple generalizations, throwing out any whose explanation involves particular facts or other nonlaws. We will be left, at the completion of this procedure, with the laws of nature.

Here are some examples that illustrate how the explanation requirement works. By this principle the fact that objects at sea level fall with an acceleration of 32 ft/sec/sec is not a law, since its truth depends on the particular fact that the Earth has a certain mass. Nor are Kepler's so-called laws really laws, since their explanations hinge not only on Newton's laws of motion but also on the fact that none of the planets is sufficiently near to one another, or sufficiently big, to significantly distort their orbits from the type of path that Kepler characterized. On the other hand, in a classical world Newtonian mechanics contains nothing but laws, since its generalizations are not explainable at all—let alone by particular facts. And consequently, Boyle's law, relating the volume and pressure of an ideal gas at constant temperature, is a law insofar as it can be explained solely by the laws of mechanics.

Of course this outline of an approach raises many questions. Nevertheless, sketchy as it is, it does enable us to clarify certain issues. A singular virtue is that it provides a rationale for excluding Hempel's example

>All objects made of pure gold have a mass of less than 100,000 kg

from the class of laws of nature. This problem has proved surprisingly difficult. First, it plainly won't help to insist that laws be general. For the generality of this example is no mere trick of syntax. Second, as we have seen, Goodman's projectibility criterion will not suffice. On the contrary, this is one of the counterexamples to his account. And third, Hempel's own proposal will not do. He suggested that the example could be excluded because, if admitted as a law, it would preclude certain phenomena that our current theories allow to be perfectly possible. But surely that rationale cannot be correct. It would dictate that we accept *no* new laws, since any new law, unless it logically followed from already accepted laws, would be bound to narrow our view of what is possible.

However, from the perspective of the 'explanation requirement', it is not hard to see why Hempel's example is not a law. That hypothesis would not be among the set of projectible, true generalizations remaining after all those whose explanation involves non-laws are thrown out. Rather, its explanation clearly would involve elements from the set of nonprojectible facts. For example, there would be a need to cite details concerning the psychological states of the controllers of huge resources, and the absence of any motive for setting about to falsify the generalization. In short, it isn't a law because it is true in virtue of particular facts.

5. Lewis's program

In reaction to the two classic difficulties in Goodman's treatment of counterfactuals—the cotenability problem and the explication of law—a radically different approach was instigated by Stalnaker (1968, 1984) and has been developed by Lewis into a broad account of temporally asymmetric phenomena. I would like to end this chapter by looking carefully at Lewis's theory. We shall find that it is faced with a variety of criticisms—avoidable, if at all, only at the cost of cumbersome, ad hoc modifications. Therefore we shall get further support for the point of view developed earlier—in which counterfactual conditionals are analyzed in terms of causation.

Lewis's theory of causation and counterfactual dependence splits into four stages. In the first place, he (1973b) says that causation consists in a chain of counterfactual dependence:

> C caused E   *if and only if* there was a sequence of events $X1$, $X2, \ldots, Xn$, such that:
>
> if C had not occurred, then $X1$ would not have occurred, if $X1$ had not occurred, then $X2$ would not have occurred, . . .
> if Xn had not occurred, then E would not have occurred

(Lewis 1987 subsequently generalizes the account to accommodate indeterministic causation.)

Second, this analysis is supplemented (1973a) with a semantic theory of the counterfactual conditional:

> If p were true, then q would be true   *if and only if*
> there is a possible world in which p and q are true that is more similar to the actual world than any possible world in which p is true and q is false

Third, Lewis (1979b) fills out the picture with an account of the features of possible worlds that make them more or less similar to actuality. The most similar worlds are said to be those in which our laws of nature are rarely violated. But exact similarity with respect to particular facts in some large region of spacetime is also a major factor and will promote similarity even at the cost of minor violations of law. A significant element in this account is that there is no built-in time bias. Concepts of temporal order are not employed at all in the principles describing the determinants of similarity.

Finally, Lewis's (1979b) fourth assumption introduces the time asymmetry that provides the ultimate basis for the directionality of counterfactual dependence and hence of causation. He makes the empirical claim that (almost) every event is grossly overdetermined by subsequent states of the world, but is not so overdetermined by its history. Or, in other words, that the future of every event contains many independent definite traces of its occurrence, although beforehand there need have been little or no conclusive indication that it would happen.

These four ingredients work together as follows. Let us imagine a hypothetical change in the course of the world—specifically, that some actual event C at time t did not occur at that time. It would be hard to reconcile this supposition with what actually happened after t, for in fact C brought about many phenomena that determine that C did occur at t. On the other hand, it would be relatively easy to square the supposition that C did not happen with the course of the world before t. Because although that history may have determined that C would occur, events are not substantially overdetermined by what preceded them. Thus at the cost of a small violation of our laws of nature, we can reconcile the nonoccurrence of C with the actual history of the world before t. But we cannot, without much greater cost, reconcile this with the actual future of the world after t. Consequently, among possible worlds without C, those that are just ours until t and then diverge are more similar to the actual world than those that are just like the actual world after t, or those that differ from the actual world before t. Thus from the asymmetry of overdetermination it follows that if the present were different from the way it is, then the future would be different, but not the past. Counterfactuals of the form, "If C had not occurred, then E would not have occurred", where C was later than E, will be false. Consequently there will normally be no chain of counterfactually dependent events leading backward in time. Therefore effects will not precede their causes.

Difficulties with Lewis's theory of causal direction emerge at each of the four stages. Let us consider some of these problems beginning with objections to the very idea that causation should be analyzed in terms of counterfactual dependence, regardless of how such dependence is itself to be construed.

1. *Causal overdetermination*. This occurs when an event is the product of more than one causal chain which would each have been sufficient to produce the event. For example, a man's death may be causally overdetermined if he is shot in the head simultaneously by two people Smith and Bloggs, acting independently of one another. In such a case the effect is not counterfactually dependent on its causes. The man would have died even if Smith had not shot him. Nevertheless, I think we would say that Smith's shot was a cause of his death. Therefore, contrary to Lewis's analysis, the presence of a chain of counterfactual dependence is not necessary for causation.

Lewis (1973b) is perfectly aware of such cases but does not regard them as counterexamples to his view. For he believes that it is unclear how to apply causal terminology to instances of overdetermination. However, even if he is right about this (which seems doubtful), it is still a mark against his analysis that it yields a *definitely* negative answer to the question of whether Smith's shot was a cause of death. For if our conception of causation neither clearly applies nor clearly fails to apply, then an accurate analysis should reflect this indeterminacy. Note that no such difficulty with overdetermination confronts our causal theory of counterfactuals. That theory will correctly *deny* that if Smith had not fired, the victim would still be alive. For, since the shots were causally independent of one another, the occurrence of Bloggs's shot will be among the circumstances in which the absence of Smith's shot is supposed.

2. *Noncausal determination*. A counterfactual dependence between events is often associated, as Lewis says, with a causal relation between them. But it need not be. There are other alternatives (Kim 1973; Sanford 1976). For example,

> If John had not been killed, his wife would not have been widowed
>
> If the last chapter had not been written, the book not have been completed

Thus, as our analysis allows, counterfactual dependence does not imply causal connection.

3. *Directionality*. Even when the counterfactual dependence of E on C *does* reflect some sort of causal connection between them, this need not be because C causes E. As we have seen, it may be, rather, that C is an *effect* of E. For example,

> If the match had lit, it would have been struck
>
> If I had jumped, there would have been a net outside the window

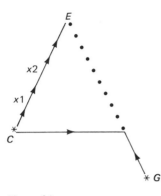

Figure 34

To handle this problem, Lewis is forced to postulate a special "back-tracking" sense of counterfactual dependence, associated with special rules for measuring the similarity of possible worlds. But I argued in section 3 that there is little pretheoretical rationale for this multiplication of senses.

4. *Causal preemption.* This takes place when the cause of an event prevents something else from causing that event. For example, Smith's shooting a man preempts Bloggs's shooting him if Bloggs is frightened off before firing by the sound of Smith's gun. If C's causing E preempts G's causing E, then, on the face of it, E is not counterfactually dependent on C because, even if C hadn't occurred, E would have been caused by G instead. Thus preemption might seem to present a problem for the counterfactual theory of causation.

But, as Lewis points out, his analysis can nevertheless be satisfied, for there may be a chain of causal dependence $(x1, x2, \ldots,)$ connecting C and E, as shown in figure 34. One might still be tempted to deny that E depends on $x2$, arguing that if $x2$ had not occurred, then neither would $x1$ nor C, so G would not have been preempted from causing E. However, says Lewis, since we are not employing 'back-tracking conditionals', $x2$'s nonoccurrence would not have made any difference to prior members of the chain; so C would still have been there and would have still preempted G's causing E.

One way of criticizing this strategy is to repeat the complaint made in point 3 regarding the alleged distinction between normal and back-tracking conditionals. Another objection emerges if we consider cases where C causes E *directly*—without there being any intermediate event X such that C causes X and X causes E. If we now suppose that C preempts G's causing E, we then have a case in which Lewis's escape will not work, and his counterfactual analysis breaks down.

Suppose, for example, that ball A rolls into ball B, causing B to move out of the way of ball D, which would have caused exactly the motion of B that A actually causes. This is a case of preemption. A's striking B causes B's motion; however, if A had not struck B, B would nevertheless have been set into motion by D. Therefore B's motion is not counterfactually dependent on A's hitting B. Moreover there are no events that mediate the causal connection between A's hitting B and B's motion. So there is no event that depends on A's hitting B and that B's motion depends on. Thus Lewis's counterfactual condition seems to be too strong.

Again, the alternative approach is not subject to this difficulty. When C preempts G from causing E, our theory will say, as it should, that if C had been absent, then E would have been caused by G. Moreover the account of causation in chapter 8 entails correctly that G is not actually a cause of E. For a cause must be *essential*, given surrounding circumstances, for the determination of its effect. But the only antecedent conditions that in combination with G will determine E are conditions that include C (or its causes or effects), and such conditions determine E without the help of G.

Admittedly, none of these arguments constitutes a knockdown argument against Lewis's approach. With enough cleverness it will no doubt be possible to save the theory from counterexamples. Indeed, Lewis (1987) does have ingenious ways of elaborating and extending his approach to deal with the problems just discussed. However, one cannot help but have the sinking feeling that we are heading for an interminable series of objections and modifications, and that even if there is an end result, it will not have the simplicity and intuitive appeal that recommended the original version. Thus, even though no irrebuttable objection to Lewis's program may be at hand, there are grounds for dissatisfaction, and reason to cast around for an alternative. These concerns are compounded as we go on to consider objections to the further stages of Lewis's theory. Let us now look at a problem that emerges when the analysis of causation is supplemented with the theory of counterfactual conditionals.

5. *Psychological implausibility*. According to Lewis, a counterfactual holds when the consequent is true in possible worlds very like our own except for the fact that the antecedent is true. But it is vital that the degree of similarity not be assessed by intuitive pretheoretical criteria. Rather, the relative importance of various factors in determining how similar some world is to our own must be retrieved from our views about which conditionals are true and which are false. For example, it has often been objected against Lewis that on his view

If the president had pressed the button, a nuclear war would have ensued

must be false, since a world in which the circuit fails and there is no war would be more like actuality than a world in which all life is destroyed. Lewis's reply, as I have indicated, is to maintain that we should infer from the truth of the conditional that the intuitive standards of similarity are not relevant. We should recognize that the appropriate standard of similarity will include something like the following ranking of how important various forms of differences are: first (most substantial), the existence of many miracles (violations of our laws), second, the absence of an exact matching of particular facts over large regions of space time, and third, the occurrence of a small number of miracles.

Now these criteria of similarity may well engender the right result in each case. However, it seems to me problematic that they have no pretheoretical plausibility and are derived solely from the need to make certain conditionals come out true and others false. For it is now quite mysterious *why* we should have evolved such a baroque notion of counterfactual dependence. Why did we not, for example, base our concept of counterfactual dependence on our ordinary notion of similarity? As long as we lack answers to these questions, it will seem extraordinary that we should have any use for the idea of counterfactual dependence, given Lewis's description of it; and so that account of our conception of the counterfactual conditional must seem psychologically unrealistic.

Finally, let us examine some further difficulties that arise when the a priori component of Lewis's theory is supplemented by the addition of his vital a posteriori hypothesis.

6. *Oversophistication.* The predominantly future orientation of counterfactual dependence, and causation, fall out as consequences of Lewis's theory only relative to a contingent, empirical assumption regarding the asymmetry of overdetermination. He assumes that given a hypothetical change in the actual course of events, it would require many miracles to preserve the actual future, but it would be relatively easy to reconcile that hypothetical change with the actual history of the world. True, some miracle would have to be supposed (assuming determinism) in order to preserve the past, but not on the scale of what would be needed to perfectly shield the future from that change.

It may seem, contrary to this assertion, that many contexts may be found in which it would be just as easy to shield the future from a hypothetical change as to shield the past. Consider, for example, the counterfactual conditional

> If his chair had been one foot to the left at 3 pm, then the rock would have hit him

The antecedent may be reconciled with our actual history before 3 pm by imagining a miraculous sudden jump in the chair's position just before 3 pm. But can we not similarly square the supposition with our actual future by imagining a miraculous sudden jump by the chair back to its original position just after 3 pm? No, says Lewis. Such a jump would not do the trick, for the chair at 3 pm in its hypothetical, temporary position emitted light waves and gravitational forces that are not exactly like the waves and forces it would have emitted if it had not been there then. Therefore, to obtain an exact match with the future, we need to imagine not only the chair jumping back but also many further miracles in order to transform the waves and forces emitted from one position into waves and forces that seem to have come from another position. But the presence of so many miracles would make for a world that is very unlike our own. That is why, if the present were different, the future would have to be different.

My quarrel with this strategy is that it is too scientifically sophisticated. We have presumably been using counterfactuals for thousands of years and have always regarded the future as counterfactually dependent on the past. It cannot be that the ground for such a view lies in the province of contemporary physics. If it were, as Lewis (1987) contends, a matter of plain observational fact that the future grossly overdetermines every event, then it would be legitimate to employ an awareness of that fact in the explanation of our linguistic behavior. But it seems to me that as things are, the fact (if, indeed, it is a fact) is fairly inaccessible—unknown to most people, even today, let alone to our ancestors. Consequently the evaluation of counterfactual conditionals cannot be conducted on the basis of such knowledge.

Lewis does attempt to provide support for his contention, but the argument is not convincing. He points out that detective stories written for the general public presuppose that crimes leave traces. However, it seems clear, in the first place, that we do not take for granted that a 'perfect' crime is impossible (although, not surprisingly, such an event would not be good material for a detective story). And, in the second place, even if clues are presented that do point unambiguously to the criminal, it is not generally supposed that his identity must be *over*determined by the clues.

7. *Empirical implausibility.* Moreover it is not at all obvious that Lewis's empirical assumption is even correct, let alone common knowledge. No doubt there is at least a grain of truth in it, provided by the fork asymmetry: the fact that correlated events have characteristic common causes but not always a typical common effect. But what Lewis needs is a very extreme version of this phenomenon. He must assume that *every* event is one of the later endpoints of a normal fork. This is not merely the trivial claim that every event has 'siblings' (i.e., other events with the same cause). It claims, in addition, that the common cause is determined by, and may be inferred from, each of the effects on its own. Lewis does not, however, give grounds for a thesis of such generality, and I see no reason to accept it.

8. *Backward causation.* Causal overdetermination is, as we noted earlier, the production of an event by more than one causal chain, each of which would have been sufficient on its own for that outcome. Now, according to Lewis, it is the nomological overdetermination of the present by the future that leads to the conclusion that if the present were different, the future would be too, which leads in turn to the future direction of causation. This idea, however, has the following counterintuitive consequence. Consider an event that happens to be very heavily causally overdetermined: for example, a collision caused by several particles simultaneously reaching the same point in space. If that collision had not occurred, then the course of history leading up to it may nevertheless have been as it actually was, but only provided that numerous miracles occur to prevent each of the particles from arriving at that spot when it did. But, as Lewis has argued in connection with the future consequences of an event, this is too high a price to pay. A closer possible world is one in which the miracles are not needed, since the recent history of the world is different and does not involve those particles moving in that way. Thus we have a past-oriented counterfactual and therefore a case of backward causation. But this is not a welcome result. Surely not every case we would normally describe as substantial causal overdetermination is really a case of backward causation!

9. *Jackson's modification.* Frank Jackson (1977) has developed a theory of counterfactuals that combines elements from both of the two major strategies we have discussed. His view is Goodmanian insofar as the truth of a counterfactual is said to depend on whether the truth of its consequent is determined through laws of nature by conditions including the truth of the antecedent. But it borrows from Lewis the way of specifying exactly what the determining condition

must be: namely, a world history just like actuality, leading up to a state of affairs containing the truth of the antecedent and otherwise as similar as possible to the actual state of the world at that time. In other words, to verify a counterfactual conditional of the form, 'If p were true (at time Ta), then q would be true (at time Tc),' simply establish that the history of the world prior to Ta and the actual state of the world at Ta—changed minimally so as to accommodate the truth of p—would determine, given laws of nature, the truth of q at Tc. Bennett (1984) has proposed a similar account, except that in his view the determining condition does not include the history of the world prior to Ta but consists solely in the minimally altered state at Ta.

Not surprisingly, this sort of idea inherits some of the difficulties of the views from which it was derived. Goodman's approach faltered on a circularity that came from supposing that the determining condition would have to be restricted to facts that would continue to obtain even if the antecedent were true. In Jackson's and Bennett's theories this problem is handled by allowing that the determining condition has to be reconciled with the truth of the antecedent, though in the most 'economical' possible way. This is tantamount to accepting Lewis's theory for just those counterfactuals whose antecedents and consequents belong to the same time slice. Therefore many of our objections to Lewis's theory will apply. Indeed, Jackson's own concluding criticism of Lewis may be turned against his own view (I have inserted the material in square brackets):

> Some similarities [to the actual world] between Ta and Tc [and *during Ta*] are important and some are not, and which are and which are not, should be part of the *output* of a theory of counterfactuals and not part of its input. Instead of decisions about which similarities in particular fact after [and during] the antecedent-time are to be preserved determining which counterfactuals with that antecedent are true, it is the other way around: the true sequential counterfactuals with that antecedent settle the similarities preserved. Hence a theory of sequential counterfactuals ought to *yield* the subsequent [and simultaneous] similarities, not draw on them the way Lewis's [and Jackson's] does. (1977, p. 8)

Moreover accounts that restrict the determining condition to occurrences no later than Ta are faced with a problem of their own. Consider a *random* event (e.g., the toss of a coin yielding heads) that takes place after Ta. Intuitively such an event belongs among the circumstances in which the antecedent is imagined to hold, so that it can be true to say, "If I had guessed heads, I would have been right". Bennett tries to handle such cases with an ad hoc modification of his analysis. But there is no natural way to obtain results of this kind within the spirit of the views just described.

This completes my discussion of Lewis's program. His account of causation began with a critical appraisal of its main competition, namely, the sort of regularity theory that I have been advocating here:

> It remains to be seen whether any regularity analysis can succeed in distinguishing genuine causes from effects, epiphenomena, and pre-empted potential causes—and whether it can succeed without falling victim to worse problems, without piling on the epicycles and without departing from the fundamental idea that causation is instantiation of regularities. I have no proof that regularity analyses are beyond repair, nor any space to review the repairs that have been tried. Suffice it to say that the prospects look dark. I think it is time to try something else. (1973b, p. 557)

It seems to me that the pendulm has swung. What Lewis said about regularity analyses is now a fair assessment of the counterfactual approach.

11
Decision

1. Decision theory in light of Newcomb's problem

You should act in some way when desirable events are to be expected if you do. Or so one might suppose, but for the following problem. The expectations of an action divide into two sorts: confidence that it would *cause* desirable events and belief that it would be highly *symptomatic* of their prior occurrence. And it may appear that only in the former case does one have a decent motive: only the causal implications matter; the merely evidential implications of your act are irrelevant to its choiceworthiness. So it seems that the initial (evidential) principle should be amended: do something if you expect it would *bring about* desirable *results*. This causal point of view has lately become orthodox opinion among philosophers of decision theory. However, I disagree with it, and my primary aim in this chapter is to explain why, and to argue in favor of the purely evidential conception of rational choice.

A second, intimately related, aim is to explain the time asymmetry of rational choice. Why do we act for the sake of desirable future events but not for the sake of the past? Obviously, if the causal theory of decision is correct, then the temporal orientation of rational choice follows immediately from the direction of causation. We act for the sake of the future because we act for the sake of events that we can cause, and those can only be in the future. However, if the evidential conception of rational choice is correct, and I think it is, then a different account of the decision time asymmetry must be provided. Given *that* conception of rational choice, the time asymmetry could exist only if past events fail to stand in the appropriate evidential relation to our decisions. We shall see, later on, how this is so.

The evidential principle, in one of its precise forms, is the requirement to maximize expected desirability. In other words, for every act under consideration, multiply the desirability of each alternative eventuality by its probability relative to the act in question, and add these products together, thus obtaining the act's expected desirability;

then perform the act with the greatest. Richard Jeffrey (1965) has long advocated a policy along these lines. On the other side, there is no consensus yet among critics of the evidential principle about a precise formulation of causal decision theory. Alternative, closely related versions have been suggested by L. J. Savage (1972), Robert Stalnaker (1980), Isaac Levi (1975) (who nevertheless does not endorse the view), Allan Gibbard and Bill Harper (1978), Nancy Cartwright (1979), Brian Skyrms (1980), David Lewis (1981), and others, each attempting to incorporate the basic intuition (that only causal consequences matter) within a rigorous, quantitative framework. For our purposes, there will be no need to become involved in the technicalities. In order to understand and adjudicate the conflict between evidential and causal decision theory, it suffices to appreciate the unsophisticated principles that motivate them: one account says that an act is choiceworthy if its performance would be *evidence* for something of value; and the other requires, in addition, that the act might be a *cause* of the desired outcome.

The dispute comes to a head in the decision context of Newcomb's problem, first reported in Robert Nozick's classic paper (1969) and given the following formulation by Howard Sobel (1979). Consider a very hypothetical situation in which you must choose whether or not to accept a sum of $1,000, no strings attached; moreover you believe that someone who is able to predict your choices (by means of a sophisticated psychological theory) will have deposited $1,000,000 in your bank account if and only if he predicted that you would decline the $1,000. What should you do? Advocates of the causal point of view must recommend acceptance of the $1,000, since it is now too late to exert any causal influence over the $1,000,000. On the other hand, according to evidential decision theory the choiceworthiness of an act depends solely on its likelihood of being associated with desirable events, and from this perspective the right choice is to decline the $1,000.

Although the general conflict between the evidential and the causal principles is sharply embodied here, we should not necessarily expect to resolve it by concentrating on Newcomb's problem. That hypothetical context is farfetched. There is no consensus about what should in fact be done in it, so it might seem to provide no intuitive evidence for or against the plausibility of the evidential principle. An apparently more promising strategy is to begin with the general question: whether the principle is in fact defective in its confounding of the desirable effects and desirable causes of one's act. If we find that such nondiscrimination engenders clear counterexamples—if it leads to the prescription of actions in *normal* situations that conflict with common sense—then the general principle must be rejected, and its particular application to Newcomb's problem is disqualified.

A well-known smoking example might appear to play just this role—the clear counterexample. Statistical data indicating a high correlation between smoking and cancer might cause your expectation of cancer to depend on your decision whether or not to smoke. And in that case the evidential principle would dictate not smoking. But now suppose you discover that the correlation results from some common genetic factor that causes both cancer and a tendency to smoke. Then it would seem silly to refrain from smoking since it's too late to avoid the dangerous gene. Yet that very course of action still appears to be required by the evidential principle, for there remains a highly expectation of cancer if one smokes. This statistical fact is merely explained, not undermined, by the new information.

2. Information screens

So the argument goes; but there is also in the literature a reply to this objection to the evidential principle, and it implicitly suggests how a whole class of similar cases may be dealt with. The central idea is that there is a reasonable inference from the statistical data: namely, that there is a high correlation between cancer and having the *inclination* to smoke. Now if you discover that this is the result of a common genetic factor, then your expectation that you have the bad gene will be entirely determined by whether you detect in yourself the inclination to smoke. If you do recognize this desire, then you might think it is likely that you have the bad gene and will get cancer. But—and this is the main point—this estimate of your chances of cancer will not be diminished in the slightest if you finally decide not to smoke. Thus, for you, smoking is not symptomatic of the bad gene, and so the evidential principles does not dictate that you should abstain.

The reasoning here is quite familiar. Suppose that events of type Z (the car radio won't work) are taken to provide evidence for the prior existence of circumstances X (the battery is defective) because it is known that X often leads to Y (no electricity enters the radio), and Y in turn frequently gives rise to Z. In that case, if we are told on some occasion that Y is present, this produces a certain degree of confidence that Y was caused by X, and this degree of confidence is unaffected by any further information as to the presence or absence of Z. In other words, knowledge of Y screens off the effect that information about Z would have had on our belief in X. In our hypothetical example it is assumed that the gene tends to cause the inclination to smoke, which tends to cause smoking. Therefore, recognition of either the presence or the absence of the inclination is an information screen: it will deprive the act itself (smoking or not) of any evidential implications regarding the gene.

In the literature on this topic one can find many other alleged clear counterexamples to the evidential principle. But they are analogous to the smoking case, and I think that they are susceptible to the same kind of rebuttal. Consider, for instance, the following extract from Gibbard and Harper (1980, pp. 165–166):

> We offer the case of Robert Jones, rising young executive of International Energy Conglomerate Incorporated. Jones and several other young executives have been competing for a very lucrative promotion. The company brass found the candidates so evenly matched that they employed a psychologist to break the tie by testing for personality qualities that lead to long-run successful performance in the corporate world. The test was administered to the candidates on Thursday. The promotion decision is made on the basis of the test and will be announced on Monday. It is now Friday. Jones learns, through a reliable company grapevine, that all the candidates have scored equally well on all factors except ruthlessness and that the promotion will go to whichever of them has scored highest on this factor, but he cannot find out which of them this is.
>
> On Friday afternoon Jones is faced with a new problem. He must decide whether or not to fire poor old John Smith who failed to meet his sales quota this month due to the death of his wife. Jones believes that Smith will come up to snuff after he gets over his loss provided that he is treated leniently, and that he can convince the brass that leniency to Smith will benefit the company. Moreover he believes that this would favorably impress the brass with his astuteness. Unfortunately, Jones has no way to contact them until after they announce the promotion on Monday.
>
> Jones knows that the ruthlessness factor of the personality test he has taken accurately predicts his behavior in just the sort of decision he now faces. Firing Smith is good evidence that he has passed the test and will get the promotion, while leniency is good evidence that he has failed the test and will not get the promotion. . . . Firing Smith would produce evidence that Jones will get his desired promotion. It seems clear, however, that to fire Smith for this reason, despite the fact that to do so would in no way help to bring about the promotion and would itself be harmful, is irrational.

But here, as in the smoking example, we can plausibly deny that the action in question would, as claimed, provide evidence for some causally prior state. Suppose Jones knows perfectly well that he is not inclined to fire Smith, and that under normal circumstances he would certainly not act in such a ruthless manner. On this basis he infers that he probably did not do very well on the ruthlessness test. And there is no reason why this estimate should change, even if he does, despite his contrary inclination, act out of character and fire the unfortunate Smith.

In each of these alleged counterexamples it turns out that some introspectible state is symptomatic of the agent's prior, value-laden condition; and his eventual act, whichever one is chosen, provides no additional information about that condition. Consequently, and contrary to our first impression, the alternate acts have no distinctive, noncausal, purely evidential implications, and so the controversial aspect of the evidential principle is not tested.

It remains, however, to examine how widely this strategy—dubbed "the tickle defense" by David Lewis—may be employed. Will there always be a convenient 'tickle' to leak to the agent advance news of what his act would otherwise reveal? If not, then the tickle defense cannot be counted on to protect the evidential principle against all alleged counterexamples. On the other hand, if so—if there always is a 'tickle'—how can we avoid the conclusion (drawn by Ellery Eells 1982) that even in Newcomb's problem a tickle will deprive the acts of any purely evidential implications, so that even there the evidential and causal principles will concur? I shall try to show that this is a false dichotomy. I shall argue that the tickle defense works in normal circumstances to safeguard the evidential principle, but that no such strategy is available in the very special conditions of Newcomb's problem.

It is not difficult to concoct highly artificial examples in which, ex hypothesi, there is no tickle, no screen, and therefore no argument for the convergence of the evidential and causal principles. For example, we could simply have stipulated that cancer be correlated with smoking, and the firing of employees be correlated with ruthlessness, *regardless of the agents' inclinations*. Alternatively, we might, following Lewis (1979a), imagine a prisoner's dilemma involving identical twins, each of whose beliefs about what the other is going to decide will be influenced by what he finds himself doing. It seems to me, however, that such scenarios do not constitute clear counterexamples to the evidential principle because they are extremely unrealistic—in exactly the same way as Newcomb's problem itself—and cannot, therefore, provide the material for authoritative intuitions.

It might seem, at first sight, that one should obviously follow the dictates of the causal theory in these cases. Intuitions may seem clearer and stronger than in Newcomb's problem. However, the reason for this is that the new cases are superficially very similar to realistic scenarios, like our initial smoking example, in which a certain course of action is undoubtedly correct. And this similarity gives that particular course of action a plausibility that, in the new context, it does not deserve. For in fact the new cases are just as far from familiar decision contexts as Newcomb's problem is. And what makes these new cases bizarre is precisely what makes Newcomb's problem bizarre: namely, the existence of evidential implications of an act that are not screened off by the agent's awareness of his own motivational state. Moreover it is no accident that we cannot find *actual* decision contexts in which acts possess such noncausal implications for their agents. Therefore it is not surprising that there is no uncontroversial counterexample to the evidential principle. Let us see why this is so.

The process of deliberation is, potentially an extremely intricate activity. It need not consist simply in the application of a decision rule to certain beliefs and desires followed by the performance of whatever act is recommended. Rather, an agent might well calculate what he ought to do, and then, in the light of his result, reassess the very beliefs and desires that entered into the calculation. This sort of process may occur repeatedly before something is finally decided on. Consequently the causal relationships between a person's antecedent physiological and psychological states and his subsequent action are often extraordinarily complex, subtle, and difficult to recognize. Those very few such causal generalizations that are manifest to ordinary observation or current psychology are bound to be crude and blatant. They certainly do not include actions resulting from relatively complex bouts of deliberation. Rather, they are restricted to simple cases where a physiological or psychological condition tends to cause an action (in a given sort of situation) by tending to cause either certain characteristic desires, or certain characteristic beliefs, or conformity to a particular decision rule, in the light of which it becomes fairly obvious what should be done. Consequently, if an agent knows that a condition CA has tended to result in action A, he can be sure that A was the product of a simple deliberation process whose salient features will be relatively easy to discern.

Often the agent will have some knowledge of the motivational differences between those who have done A and those who have not. He may know, for example, that the beliefs of the individuals were fairly uniform and determined in familiar ways by observation, hearsay, and so on. And he may know that the operative distinction between the individuals was a difference in their desires. Then he might well be able to determine what desires would have led to the performance of A, and having done this, he can see by introspection whether or not he has them. So in many cases the agent's awareness of his desires will constitute a tickle that screens out the epistemological import of his subsequent act.

One should not object here that a person's desires may not always be accessible to introspection. This is true but irrelevant. Our screening defense needs to be employed only for situations that provide alleged counterexamples to the evidential principle. And there can be a counterexample to the evidential principle only if the principle is applied, and therefore only if the beliefs and desires of the agent *are* known by him at the time of deliberation.

Of course it will not always be possible to identify a screening factor so easily. Nevertheless, the following general method, should work in all normal circumstances. Suppose the agent, on learning certain statistical information, begins his process of deliberation holding that his act, A or $-A$, will be symptomatic of the presence or absence of a certain value-laden antecedent state CA. That is,

$$P(CA/A) = e \quad \& \quad P(CA/-A) = f \quad \& \quad e > f$$

He believes that in his own case the causal process producing whatever act he will perform will include an activity of deliberation in which his desires and beliefs (including his belief in the evidential relevance of his act) are employed in the evidential principle to generate a rational recommendation. Since his belief in the evidential relevance of his act is derived straightforwardly from statistical information, he must regard himself as a representative member of the group of individuals about whom the statistics were compiled. Therefore, he must believe that *deliberation* was involved in the production of the acts in the group. To repeat, if he does not assume this, then he should not believe in the evidential relevance of his act, since he would not be taking himself to be representative of the group. Now, I have argued earlier that for correlations between physiological/psychological states and acts to be discernible to us, they must be mediated by relatively *simple* deliberations. Therefore our agent may assume that those people in the sample who did A (or $-A$) did so, not as the product of some elaborate sequence of continually revised calculations, but after a relatively straightforward deliberation yielding A (or $-A$) as the clearly recommended act. That is to say, he can take the correlation between CA and A to be a reflection of the tendency of CA to cause a person's *initial* deliberation to yield the recommendation that A be done. This now gives rise to the following way for the agent to identify an epistemological screen. Let him actually conduct an initial stage of deliberation. Let him use his beliefs, his desires, and the evidential principle to determine which act would be rational. Having done this, he will then be aware of a state of his own mind—namely, believing A (or $-A$) to be the initially recommended act—that he can assume was present in the minds of all those in the statistical sample who did A (or $-A$). In other words, call the state of believing A ($-A$) to be the initially recommended act, BA ($B-A$), then the agent will hold

$$P(CA/A \& BA) = P(CA/-A \& BA) = e$$

and

$$P(CA/A \& B-A) = P(CA/-A \& B-A) = f$$

Thus, on recognition of the epistemological screen, BA or $B-A$, his initial belief in the noncausal, evidential relevance of his act will become extinguished.

One might object that although such a screen might be available to anyone prepared to go through the fairly complex process of identifying it, there will be cases in which the agent is naive, lazy, or must make a quick decision and does not have the time or ability to do this. Consequently his act may retain noncausal implications, and he may be led, by the evidential principle, to an act that we would judge to be irrational. However, this criticism neglects a certain systematic equivocation in the evaluation of actions. Actions are always judged in relation to desires and beliefs that are themselves susceptible to evaluation. Therefore an act may be criticized as irrational because it was based on irrational beliefs, even though it was correct relative to those beliefs. In particular, the irrationality of someone's past-oriented action may be attributed to his irrational belief that the act was evidentially relevant to the past, and not blamed on his use of the evidential principle. Moreover the preceding argument for the evidential irrelevance of acts is fairly general. So there would be no need for an agent to repeat it on every occasion. A reasonable assumption in normal circumstances, especially if one is pressed for time, is that one's acts do not have non-causal implications. Indeed, it would be irrational to assume the contrary. For these reasons, if someone does not correctly evaluate the significance of his act, we may judge it to be irrational without thereby impugning the evidential principle.

However, the screen strategies will work only for decisions involving the simplest deliberations. They will not work in general—in unrealistic, hypothetical cases—and they certainly cannot be relied on in Newcomb's problem. To see this, note, first of all, that in Newcomb's problem the causal chain leading to a given act is not mediated by any particular, easily recognizible, introspectible state. No characteristic desires or beliefs lead to one or another of the acts, so no such state can be employed as an epistemological screen. Moreover, if someone performs an initial deliberation, employing certain provisionally accepted beliefs, desires, and decision rules, and reaches the conclusion that taking the $1,000 is best, he cannot then conclude that there would now be nothing more to be learned from actually doing it. The known correlation is between A *itself* and CA, and he has no right to assume that doers of A were precisely those people who discovered in their *initial* deliberation that A is best. Rather, doers of A will include many people who achieved that result by the most contorted of deliberations.

My point here is not simply that we oughtn't to expect a tickle in Newcomb's problem, but rather that the problem should be formulated in such a way as to make it clear that this is impossible. Only given such a formulation will the problem be philosophically interesting, for only then will it drive a wedge between the evidential and causal principles. No doubt we can make up variants of Newcomb's decision context in which there *is* a tickle indicating the physiological condition that tends to produce both refusal of the $1,000 and the prediction to that effect. Both decision theories will then recommend accepting the $1,000, and on those occasions when the tickle is present, the prediction will be falsified, the subject will get the $1,000 and will find himself with $1,000,000 in the bank. However, it remains possible to characterize a different, and more important, version of Newcomb's problem. Here, the predictor's psychological theory is perfectly adequate, and any mistakes derive solely from calculating errors that are not associated with paricular types of subject. In such a decision context there can be no epistemological screening, and so the evidential and causal principles will dictate opposite policies.

3. Revisionism by Eells and Jeffrey

Contrary to this position, Ellery Eells (1982) has argued that screening always takes place, that rational acts are never evidence (to the agent) of causally prior states, and therefore that the evidential and causal principles never diverge from one another, even in Newcomb's problem. Let us examine his argument.

He begins with the following assumptions about the agent's state of mind just prior to the moment of decision.

1. The agent knows what his beliefs and desires will be at the time of decision. That is,

$$P(M) = 1$$

where M accurately describes his beliefs and desires and P represents his subjective probability assignments (degrees of belief).

2. The agent knows that he will perform a given act A just in case (DA) he rationally determines that this act is correct. That is,

$$P(A \longleftrightarrow DA) = 1$$

3. The agent's beliefs and desires screen off any probabilistic dependence between what he will rationally determine to be correct and any potential physiological cause of action, CA. That is,

$$P(DA/M \& CA) = P(DA/M \& -CA)$$

Now from 2 and 3, Eells infers

4. $P(A/M \& CA) = P(A/M \& -CA)$

from 1 and 4 he obtains

5. $P(CA/A) = P(CA/-A)$

Thus the agent's performance of action A would not constitute evidence for or against the presence of any potential causal antecedent CA of A. Now there is presumably a certain invariable causal antecedent of the decision to take the $1,000—an antecedent that the predictor is able to recognize as such and that determines his action with respect to the $1,000,000. However, from statement 5 it follows that the act of taking the $1,000 will have no implications regarding that antecedent and therefore no implications regarding the probability of the predictor's having decided to give the $1,000,000. Thus the evidential principle dictates, in agreement with the causal principle, that one should take the $1,000.

There is no doubt that this argument is valid: statement 5, as formalized, does indeed follow from the formalized versions of 1, 2, and 3. However, the premises of Eells's argument are far from uncontroversial. Statement 3, in particular, is quite implausible and the argument given in favor of it is quite unpersuasive. We may grant that if M mediates the causal chain between CA and A, then, construed as an *empirical propensity* or relative frequency thesis, 3 would be correct. However, Eells's argument requires that the probabilities in 3 be *subjective* probabilities, for deliberations are based on beliefs, not unknown facts; yet 3 construed in that way does not follow from 3 construed as a propensity thesis.

Suppose we have established that one of two processes is going on. But we have no idea which one, and no knowledge of these processes except for each of their initial and final states. Observing an intermediate state may not help us tell which process we are witnessing and what the outcome will be—even though we are aware that the objective propensities of the two possible results, given the observed state, are independent of what preceeded it. Similarly, in the case of decision, we may know that M describes a mediating state without being able to give it the slightest epistemological significance and without drawing any conclusions about the probability of CA from its discovery.

In response to this criticism it might be said (as Eells does in 1982, p. 164) that an agent *should* be able to draw inferences from his discovery of M about the probabilities of the alternative acts. That he can do this should follow from our joint assumption that he will act rationally and that the beliefs and desires of an agent determine which of his acts would be rational. However, this response presupposes that what is meant by a *probability relative to M* is not simply a *probability relative to M's discovery* but, rather, a *probability relative to M being interpreted* (i.e., a probability relative to M being discovered and reasoned about to the extent that its consequences for action are inferred). And we can question whether this latter probability is the appropriate one for use in deliberation. The reason for concern about this is that when the expected utilities of alternate acts are calculated for the purposes of deliberation, it is obviously required that the agent does not yet know which act he will perform. If he did already know that he was going to do A rather than B, then further deliberation would be pointless. Moreover the expected utility of B could not be calculated since it would involve conditional probabilities relative to a condition *known* to be false. Now, given premise 2—namely, the agent knows he will do what he determines is rational—it follows that as soon as M has been apprehended and interpreted, the agent knows what he is going to do. Thus Eells is faced with a fatal dilemma. Either he supposes that the probabilities in 3 are to be taken as *relative to M being interpreted*—in which case those probabilities are irrelevant to deliberation, since they are relative to a knowledge of what will be decided, or they are not to be taken in this way—in which case premise 3 must be rejected. The only defense of 3, in its required sense, is to maintain that given the discovery and interpretation of M, all that can be definitely inferred is which act is rational, and to maintain that this does not constitute knowledge of which act will be performed. For, despite one's best intentions to act rationally, something could intervene to produce a different act. However, such a response is of no help to the argument; it props up 3 at the cost of abandoning premise 2.

A less formal way of putting the difficulty is this. No matter how long the process of deliberation continues, a reflective agent in Newcomb's context can never draw definite conclusions from his mental state about what he will do until after the deliberation is over and done with—at which point of course it is too late to use that knowledge in the deliberation.

Inspired by Eells's analysis, Richard Jeffrey (1981, 1983) has proposed a modified version of his evidential theory, called "ratificationism," which is specifically designed to deprive noncausal implications of motivational significance. The idea is that one should choose act A, only if one recognizes that even after A has been finally chosen (but before it is actually performed), A will still have no less expected value than any of the alternative acts. In other words, A is ratifiable, and thereby choiceworthy, if and only if the expected value of A, relative to the supposition that A will be finally chosen, is no less than the expected values of any of the alternative acts, again relative to the supposition that A will be finally chosen. Now, Jeffrey does not embrace Eells's premise 2. Consequently there is no danger that probabilities relative to the choice of one act and the performance of a different act will not be well defined. However, in avoiding this problem, another one is created. Having been careful to distinguish between, on the one hand, *finally choosing an act* and, on the other hand, *actually performing* it, and having opened up a causal gap between them, it becomes natural to regard the decision problem as concerning just the former, and to regard the latter event as simply a very probable consequence of what is directly chosen. By not treating the matter in this way, ratificationism becomes susceptible to predictable forms of counterexample. Suppose that, for whatever reason, the causal consequences of finally choosing A but nevertheless doing B are known to be much worse than choosing and doing B. In that case A might well be ratifiable, even though it is obviously wrong. For example, imagine you are guaranteed a system of rewards according to the following rule:

> Choose A; do $A \rightarrow \$50$
> Choose A; do $B \rightarrow \$0$
> Choose B; do $A \rightarrow \$150$
> Choose B; do $B \rightarrow \$100$

Supposing that A is finally chosen, the expected values, from that perspective, of doing A and doing B are \$50 and \$0, respectively. Thus A is ratifiable even though any rational person would choose, and probably do, B, which, as it turns out, is not ratifiable. Jeffrey is aware of such difficulties with ratificationism and has now returned to his original (1965) interpretation of evidential decision theory.

Although I have argued against the detailed proposals of Eells and Jeffrey, I certainly accept that their basic insight contains a great deal of truth: namely, that the process of deliberation may provide the agent with a tickle—(an epistemological screen that destroys the symptomatic implications of his act). This point is indeed essential in the defense of the evidential principle against alleged counterexamples like the smoking case, but I have indicated earlier the alternative way in which I think that this defense should be conducted.

4. Four arguments against the causal theory

Let us summarize our position so far. The critics maintain that the evidential principle is clearly wrong in allowing the desirability of an action to depend on its purely evidential implications as well as its causal implications. However, the alleged counterexamples designed to establish this criticism fail to do so. For it turns out that in all realistic cases (though not in all conceivable cases) in which an act may seem to have noncausal implications, there is an introspectible state that provides advance information regarding those implications and deprives the act itself of diagnostic significance. So the controversial aspect of the evidential principle is not tested. Nor is Newcomb's decision context any help here. For although it does have the structure needed for a test case—contrary to Eells's argument, the two choices do have different evidential implications—the trouble is that our intuitions about what should be done are unsettled. A genuine test of the principle requires a case where, on the one hand, the alternate acts do have variant evidential implications and, on the other hand, our considered intuitions are firm and uncontroversial. In the absence of such an example, we are left with two competing rules: the evidential one, and the causal principle, where the desirability of an act depends only on its expected causal consequences. Each of these give intuitively right results for all the clear-cut cases we have thought of. But they conflict in Newcomb's problem. Can some reason be found to choose between them?

One fair possibility is that the notion of 'rational act' is a *cluster* concept (Putnam 1962) involving more than one application criterion. Repeating our example from the discussion of the direction of causation in chapter 8, consider the classical idea of 'straight line'. This term had many application criteria, including 'satisfying the axioms of Euclidean geometry', 'being the possible path of a light ray', and 'being the possible path of a freely moving particle', which were all assumed to be consonant with one another. None could have been singled out in advance as having a priviledged status—as representing the *definition* of the notion. Only in the context of a new theory, in which these criteria are no longer equivalent, could it be decided, on grounds of theoretical simplicity, which criterion to continue to associate with the term "straight line".

As Wittgenstein emphasized, language is a practical instrument, and it evolves in such a way as to be useful in our actual environment, for dealing with situations that actually arise. Therefore it should not be surprising that when language is confronted with an extremely bizarre hypothetical circumstance, there may be no clear way to apply it. Nor should we expect, even if some particular way of proceeding is the correct way, that this fact is somehow implicit all along in our earlier linguistic practice.

Applying these considerations to the problem at hand would suggest that there is no psychological fact determining which choice in Newcomb's problem conforms to our conception of rationality. According to this view, our idea of rational act is equally associated with both the evidential and the causal criteria since circumstances in which their dictates would diverge were not anticipated in the formation of that notion. Thus what to say about such circumstances is not determined by our present idea but involves some extension of it. Such an extension need not be a matter of arbitrary convention, however. As in the case of 'straight line', our decision may be the product of rational considerations.

If 'rational act' is a cluster concept, and if just one of the extensionally adequate criteria is to survive the filter of Newcomb's problem, then it seems to me that the evidential theory is the more plausible candidate. I have four reasons for this choice. Let me state them in what I take to be the order of increasing strength.

1. *Complexity.*   The evidential criterion is simpler than the causal criterion and is therefore to be preferred on grounds of theoretical economy. To see this let us look more carefully at the difference between them.

The evidential criterion is a uniform rule. There are no conditions or qualifications in its requirement to assess the contribution of *act A*'s *value given state S* toward *the total value of A* by multiplying it by *the probability of S given A*. Thus we have the absolutely general rule

$$V(A) = P(S1/A) \cdot V(A/S1) + P(S2/A) \cdot V(A/S2) + \ldots$$

where $S1$, $S2$, $\ldots$, is a finite sequence of exhaustive and mutually exclusive states.

The causal criterion, on the other hand, is not a uniform rule. To apply it, we must first establish a way of dividing up the states $S1$, $S2, \ldots$, into those parts, $K1$, $K2, \ldots$, that are causally independent of the choice, and those parts, $C1$, $C2, \ldots$, that are not. Thus perhaps $S1 = K2 \& C5$, $S2 = K3 \& C1$, $S3 = K2 \& C1$, and so on. To calculate A's value given each of the possible fixed circumstances (i.e., to obtain $V(A/K1)$, $V(A/K2), \ldots$), we employ the evidential rule, since the remaining alternative factors $C1$, $C2, \ldots$, are causally dependent on the choice. That is,

$$V(A/K1) = P(C1/A \& K1) \cdot V(A/C1 \& K1) + P(C2/A \& K1) \cdot \\ V(A/C2 \& K1) + \ldots$$

$$V(A/K2) = P(C1/A \& K2) \cdot V(A/C1 \& K2) + P(C2/A \& K2) \cdot \\ V(A/C2 \& K2) + \ldots$$

Now, to get $V(A)$ from $V(A/K1)$, $V(A/K2), \ldots$, we are to use a different rule:

$$V(A) = P(K1) \cdot V(A/K1) + P(K2) \cdot V(A/K2) + \ldots$$

where the alternative values are not weighted by their *conditional* probabilities relative to the acts, because $K1$, $K2, \ldots$, are not causally dependent on the acts. Thus by ordinary intuitive standards the causal criterion is more complex.

The preceding formulation of causal decision theory is given by Skyrms (1980), and a similar version has been developed by Lewis (1981). However, we should acknowledge that there exists an alternative formulation of the causal theory, due to Gibbard and Harper (1978), that appears on the surface to be as simple as the evidential rule. Indeed, it is just like the evidential rule except for employing the probabilities of counterfactual conditionals instead of conditional probabilities:

$$V(A) = P(A \;\square\!\!\rightarrow S1) \cdot V(A/S1) + P(A \;\square\!\!\rightarrow S2) \cdot V(A/S2) + \ldots$$

However, the simplicity of this rule is an illusion, much like the apparent simplicity of Goodman's "grue" hypothesis. For, in order to make sure that the rule conforms to intuitive judgments, it proves necessary to adopt the following principle.

If state Sn is believed (to degree 1) to be causally independent of act A, then $P(A \;\square\!\!\rightarrow Sn) = P(Sn)$

Thus the Gibbard and Harper rule requires implicitly what the Skyrms and Lewis rules require explicitly: namely, that causal considerations be invoked in calculating the expected value of an act. Insofar as it does without the concept of causation, the evidential principle is simpler and preferable. (See Eells 1982, p. 147, for discussion of this point.) Notice that my argument here is not of the familiar scientific kind where one hypothesis is said to be more *plausible* than other because it is simpler. Rather, we are concerned with how our cluster concept of rational act would respond to the impact of Newcomb's problem. I am suggesting that the relative conceptual simplicy of the evidential rule provides some reason to think that it, rather than the causal rule, would survive.

2. *Guaranteed irrationality.*   A serious difficulty in causal decision theory has recently been brought to light by Reed Richter (1984): namely, that there are circumstances in which every single one of an agent's choices will be branded by the causal rule as irrational. This problem arises because the agent's beliefs—those he employs in his calculation of what to do—may be undermined by the result of his deliberation, so that this result is itself undercut. In other words, it can happen that, having decided what to do, the agent is led to revise some of the beliefs that entered into his decision, and revise them in such a way that the initial decision is shown to be wrong. And this could occur for each of the agent's possible choices.

There is already a hint of this difficulty in the context of Newcomb's problem. According to causal decision theory, I am supposed to begin with some guess about whether the million is there or not; I then calculate that I should accept the $1,000; and then I reduce my initial estimate of the chances of getting the million. However, in this case the revision doesn't really matter, since, even on the basis of my new estimate, the same choice as before is dictated. But now let us consider a modification of Newcomb's problem, which I have designed so that the revision will cause trouble. Imagine the agent knows he has a sympathetic benefactor who is likely to compensate him should worse come to worst in his Newcomb choice. Specifically, he believes that if he declines the $1,000 and finds nevertheless, that the million is not there (i.e., he seems to be headed for nothing), then his benefactor will probably give him $10,000. What should he do? Following the causal theory, a decision to take the thousand will indicate that the million is absent, so that the best thing would be to aim for the compensation and decline the thousand. On the other hand, a decision to refuse the thousand will suggest that the million is present, and that an extra thousand would be gained by the opposite choice.

Thus he can neither take it nor leave it—both decisions are irrational.

Note that what gives rise to this absurdity is (as we saw in our discussion of complexity) that beliefs enter in *two* ways into the causal theorist's calculation. First, there are the unconditional beliefs that the agent has about the circumstances he takes to be beyond his control, and second, there are his conditional beliefs about the probable consequences of his choice, given assumptions concerning the uncontrollable factors. The problem arises because the first collection of beliefs can be made to seem implausible in light of the choice they helped to determine. However, no such difficulty could plague the evidential account. That rule employs only conditional beliefs—those relative to the agent's options. Such beliefs cannot be affected by the arrival at a decision.

3. *Arbitrary time bias.*   My third reason for advocating the evidential point of view is that the causal criterion embodies an arbitrary time bias. It is fair to ask, I think, *why* we should insist that the desirable states for the sake of which we are acting be *effects* of our actions and not be *causes* of those actions. It seems to me that this restriction looks particularly arbitrary and unreasonable when we examine what is fundamentally responsible for this highly prized causal directionality. When two events are causally connected, what makes one of them the cause and the other the effect? Perhaps the most natural answer to this question is that causal directionality derives somehow from temporal directionality. Very roughly speaking (see chapter 8 for details) the earlier of the two causally connected events is constitutively, the cause and the later one the effect. If this answer is accepted, it poses a difficulty for the causal criterion. For, it follows from this answer that a requirement that rational acts be for the sake of what they might cause, and not what they might be noncausal evidence for, is simply a consequence of a requirement to act for the sake of the future and not for the sake of the past. But why impose this temporal bias? It seems just as arbitrary as a requirement that rational acts be for the sake of the East and not the West.

In response to this objection a causal decision theorist has two options, both unpromising. One strategy is to admit the time bias but deny the accusation of arbitrariness—to contend that the sheer temporal order of desired event and possible act is indeed relevant to the act's choiceworthiness. To sustain this view, it would be necessary not only to identify some anisotropy (some intrinsic asymmetry) within time itself, but also to explain the motivational relevance of that anisotropic feature. Prospects here look dim.

Another possible line of defense would be to reject the temporal account of causal priority. But then we need a replacement, and none of the alternatives entertained in the literature (either singly or in combination) is especially helpful to the causal theorist. Some just smuggle in a time bias through the back door (e.g., Mackie's 1974 strategy of defining causation in terms of a notion of 'fixity' that applies with conceptual necessity to all past events but not to all future events). Some derive causal directionality from temporally asymmetric physical phenomena (e.g., entropy growth, according to Reichenbach 1956; gross overdetermination of the past by the future, according to Lewis 1979b) whose bearing on rational motivation is just as obscure as the bearing of time order. And, finally, there is the idea (see Gasking 1955; von Wright 1971; and Healey 1983) that the concept of causation is somehow derived from our notion of rational action. Thus it might be that the word "causation" is partially defined by the principle

> Causation is the relation between our decisions and the events for the sake of which they are made.

But this strategy is clearly no use to causal decision theory since, in the first place, it implies that we have a conception of rational decision that is prior to, and therefore independent of, our notion of causation; and second, it provides a plausible alternative to the causal theorist's explanation of the apparently intimate association between the direction of causation and the usual orientation of deliberation. Thus causal decision theory does not square with any explanation of causal priority.

4. *Normative inconsistency.* Perhaps most important, let us not neglect the *normative* character of the concept under consideration. In particular, we should not forget the intimate relationship between assertions about what someone ought to do and desires (of a certain sort) concerning what he will do. Indeed, the connection is so close that some philosophers have been tempted to maintain that normative assertions do not really purport to describe facts, as their grammatical form would suggest, but are really just expressions of taste—emotional outbursts which can be neither true nor false.

Without going quite so far, we can nevertheless say that, whether or not there is any descriptive content in a normative statement that X ought to be done, its illocutionary force is partly to express a *pro attitude*—a certain kind of desire—to the effect that X be done. In that case I think it would be paradoxical to suggest that one might have this desire and also think it rational to *hope* that X will not be done. However, advocates of the causal theory are committed to just this inconsistency. For, although they recommend taking the $1,000 in Newcomb's situation, they also recommend attempting to make oneself into the sort of person who will decline it, and they have no complaint against someone who hopes he has achieved this state. You should hope, for his sake, that a friend in this situation would decline, and be happy when this hope is realized. But at the same time, say the causal theorists, you must assert that he ought not to refuse it. This means that you simultaneously have a proattitude toward the agent's hoping to do X rather than Y, and a proattitude toward Y rather than X actually being done. However, since what the agent prefers to do and what he actually does are reliably correlated with one another, one would think that the first proattitude would imply the opposite of the second, and vice versa. Thus it seems that causal decision theory is embroiled in a contradiction that stems from neglecting the essential relationship between normative assertions and expressions of desire.

Against this criticism of the causal theory one might be tempted to maintain that the rational hope is not, as I have claimed, that the $1,000 be declined. Rather, one might think, it is most reasonable to have the compound hope that the $1,000,000 is there and that the $1,000 is accepted. In response, I would say that we must always consider such hopes in relation to the alternatives. If one hopes for X rather than $-X$, one may simultaneously and consistently hope for $-X \& Z$ rather than any of the other three logically possible combinations, $X \& -Z$, $X \& Z$, and $-X \& -Z$. In particular, the preceding compound hope, which I agree is rational (relative to the three natural alternatives), does not at all conflict with the hope, also rational, that one will reject, rather than accept, the $1,000. In the same way, supposing your doctor says that if certain test results show that you probably have a dangerous disease, he will send you $150 instead of a bill; then you are surely not hoping to find money as you open the envelope. You may well hope for misleading results, so that you are healthy and still get the $150. Nevertheless, you will not be unhappy to see the bill.

An alternative reaction to the criticism would be to complain that it just begs the question: that it takes the essence of the causal theory— which is to *separate* the motivational value of an action (its choice-worthiness) from the 'news value' of an action (how much we should hope it is performed)—and simply labels this distinction paradoxical. My strategy, however, is not merely to dismiss the distinction out of hand but to argue that it conflicts with the expressive nature of normative assertions. I am suggesting that to hold that something ought to be done is, in part, to have a certain favorable, recommending attitude toward its being done; and I am suggesting moreover that it is incoherent to have this attitude toward something without also having it toward known accompaniments. In particular, one cannot coherently have this attitude toward X rather than Y being done, and toward Y rather than X being preferred by the agent. Clearly it would require a considerable space to establish these theses properly. All I have been able to do here is to note their plausibility, respond to a couple of likely objections, and draw some conclusions to the detriment of the causal theory.

Thus, although the notion of rational choice may at present be a cluster concept, the causal principle is a relatively insecure element that should not, and would not, survive the competition engendered by a confrontation with Newcomb's problem. Admittedly none of our four considerations is a knockdown argument for this. But they surely have some force and, in the absence of anything to be said on the other side, perhaps enough force to justify the conclusion: to embrace evidential decision theory, decline the $1000 in Newcomb's problem, find $1,000,000 in the bank, and obtain a happy convergence of reason and well-being.

5. The value asymmetry

As Derek Parfit (1984) has recently emphasized, there is a sense in which we care more about the future than the past. We would like there to be good times ahead of us; on the other hand, we feel relieved when unpleasant things are over and done with. One way of putting the attitude—no doubt exaggerating it—is to say that if we were now given the power to locate in our lives various desirable and undesirable experiences, we would choose to have the pleasant ones in the future and the bad ones in the past. Another way of expressing the point is to say that the values of certain future events are magnified compared to the values of similar past events. What this means, for example, is that something desirable, like having a good meal tomorrow, has (say) twice the positive value of having had a good meal yesterday; and something undesirable, like stubbing one's toe, also has twice the value (in this case negative) as having stubbed it yesterday.

But is it true that we simply care more about the future? One defect in this characterization of the phenomenon is that it is too broad. Consider, for example, the terrible fact that millions don't have enough to eat. Surely it is no worse that a particular child, sometime in the future, will starve than that a certain child, sometime in the past, should have starved. Reflection on such cases suggests that it is not generally true that the values of future events are inflated. Rather, it seems that the inflation is restricted to our *own* experiences. We care more about our own future pleasures and pains than we do about past ones. Or to put the point more abstractly, we now desire that our own future selfish desires will be satisfied, but we do not presently desire, to the same degree, that our past selfish desires were satisfied.

Why should this be? To begin with, let us separate the problem of explanation from that of justification—the question *why* the asymmetry exists, as opposed to the question *what rational basis* do we have for this attitude. Of course, these questions may, in principle, have the same answer. It could be that the attitude exists because we recognize its reasonableness. But we can't simply assume that this is so. On the contrary, I suspect that the source of the asymmetry has nothing to do with considerations of rationality.

Rather, it is plausible to explain the asymmetry, as Parfit suggests, in evolutionary terms. One would expect that it has selectional value. That is to say, an organism that wanted its future selfish desires to be satisfied would flourish relative to an organism that didn't care; however, there is no particular advantage in wanting past desires to have been satisfied. In order to sustain this explanation, it is important to show:

1. That having the desire that future selfish desires be satisfied would increase the chances that those future desires will be satisfied.
2. That the satisfaction of future desires would be conducive to survival and reproductive success.
3. That having the desire that *past* desires were satisfied would not increase the chances that those past desires were satisfied.

And these asssumptions are eminently plausible. Point 1 is a restriction of the generalization: desiring a future event increases the chances of its occurrence. Point 2 follows from the fact that food, warmth, sex, avoidance of injury, and so on, are what we will selfishly desire and are also conducive to survival and reproduction. And point 3 reflects the future orientation of the process—desire, deliberation, decision, action, and fulfillment. This is the mechanism by which desire and fulfillment can be correlated, but no such mechanism exists to associate past-oriented desires with their fulfillment.

Of course, despite its attractiveness, this is an *empirical* explanation, and it could be false. In particular, we must beware of the fallacy (emphasized by Gould and Lewontin 1979) of assuming that every useful trait of an organism has been selected for. Also our explanation, even if true, could well be substantially improved. For example,

I have started with the fact that the trait 'desiring that future selfish desires be satisfied' would not be most efficiently achieved by selection for 'desiring that *all* selfish desires be satisfied'. We would like to know why this is so. But evidently any such deeper explanation of the value asymmetry calls for rigorous work in evolutionary biology.

In conclusion, no doubt we do care to some extent about our past experiences because of their implications for the future and because we enjoy pleasant thoughts and images of them. Some of our regard for future pleasure comes from similar sources. However, it is plausible to suppose that the predominant factor—that which introduces the substantial time asymmetry—is the special reproductive advantage of concern for the future.

12
Conclusion

The 'mystery of time' grows out of the convoluted interaction between time and our network of basic concepts. There is a sense that only something deeply enigmatic could play such a peculiar role. In this book I have tried to combat that impression. I have proposed ways of explaining a wide range of temporal asymmetries, and have thereby sought to show that time is not essentially unfathomable.

There is, I think, a single physical fact on which all the asymmetries depend, in one way or another. This fact (which I called "the fork asymmetry") is that whenever two event types are correlated with one another, they are embedded in a V-shaped chain of nomological determination, but need not be embedded in a Λ-shaped pattern; thus there is always a characteristic antecedent event, but there need be no characteristic subsequent event. In chapter 4 I describe this phenomenon and speculate about which cosmological conditions might account for it. But this is really the business of physics. My main concern has been to examine its philosophical ramifications, and in the rest of the book I try to show that all the asymmetries under consideration may be traced back to it. The rough idea—*very* rough—is that the fork asymmetry leads to the contrast between our knowledge of the past and ignorance of the future and that this epistemological asymmetry has two important consequences. First, it fosters the idea that the past is in some sense 'more basic' than the future, and thereby inclines us to *explain* the future in terms of the past, and not vice versa. Second, it implies that rational action (since it involves a process of *discovery*, during deliberation, of what is to be done) must be oriented toward the future—which explains why we *care* more about the future than the past.

This attempt to cram our conclusions into a nutshell leaves out many vital qualifications and important parts of the story. But time asymmetry turns out to be a messy affair, and the results of clearing it up cannot, unfortunately, be captured in a single resounding thesis. In particular, I reject the idea that some 'directionality' inherent in time itself is the source of temporally asymmetric phenomena. Indeed, there is no reason to think that time itself is in any way asymmetric. Nor do I believe that there is any other uniform explanation of the asymmetries. To mention one possible candidate, it is clear that they cannot be attributed simply to conventions of language. It might seem initially reasonable to suppose that *some* of the asymmetries—for example, the typical time order of cause and effect—are products of stipulation. However, there are others—such as the prevalence of decay and our striking knowledge of the past—that are, unquestionably, solid empirical facts. And in that case the plausibility of relating all the asymmetries to one another (not to mention the existence of Quine's general arguments against a priori truth) quickly removes the temptation to regard any of the asymmetries as mere matters of convention.

Thus no overall 'philosophy of time' is contained in this work. Instead, what we are left with is a series of particular explanations of the individual asymmetries and a map of their interrelationships. True, I have claimed, and argued for the view, that the asymmetries are profoundly unified and should be treated as a whole. But this position is not especially controversial. The main content of this book is in the specific arguments and ideas that are deployed to treat the problems of causation, time travel, decision theory, counterfactuals, and so on, and in the particular way that the alleged solutions are intergrated with one another.

At the very least, I hope to have confirmed that the cluster of issues involving time asymmetry forms a field of research worth pursuing. It is valuable not only because time is such a fundamental and ubiquitous concept but also because we get the opportunity to examine a large network of important concepts from a single perspective. And this focus can be very fruitful. For, although no element of the network may be understood in isolation from the other elements, a simultaneous assault on them all looks overwhelmingly difficult. However, by concentrating on their temporal relations, we confine ourselves to a manageable problem whose solution can provide a key to the overall picture, and thereby to a full understanding of each individual concept.

Each of the previous chapters has been about a particular piece of the global problem—a particular asymmetry in time. To conclude this essay, I would like to return to the whole picture, summarizing the main conclusions and giving some indication of where further work is especially needed.

Let me begin by repeating the explanatory map (figure 35), which was given in chapter 1, and for which I have been arguing in the

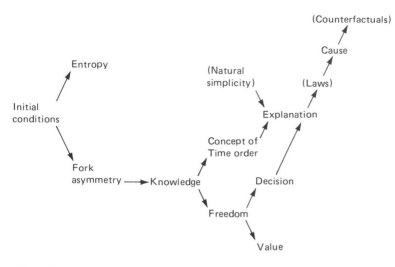

Figure 35

course of this essay. As might be expected, the story begins with the origin of the universe (or, at least, the origin of the current phase of expansion). In particular, two features of the initial conditions are important: first, the fact that there was an uneven distribution of energy (high macroscopic order) and, second, the fact that the conditions were otherwise chaotic (the highest possible degree of microscopic disorder consistent with the macroscopic order). These circumstances help to explain the (approximate) second law of thermodynamics: that is, why we encounter many processes, such as mixing and decay, in which entropy increases and almost none in which entropy spontaneously declines. In addition the 'initial micro-chaos' condition explains the phenomenon I called "V-correlation": that correlated event types are invariably associated with some characteristic *antecedent* event within a V-shaped pattern of nomological determination. For, if there were no such connection between some pair of associated event types, then their correlation could be traced back to a correlation in the initial conditions, which would violate the 'initial chaos' assumption. The fork asymmetry consists in the truth of V-correlation in conjunction with the falsity of its time reverse. For there is no principle implying that correlated events tend to be associated with a characteristic *subsequent* event.

Bearing in mind the fork asymmetry, one is in a position to see why it is that we know more about the past than we do about the future. For the processes that give us information about the past are typically instances of the V-correlation principle. Recording systems tend to cluster into a small proportion of their permissible states, and this clustering constitutes a striking correlation—a coincidence that is explained away by the prior occurrence of characteristic antecedents. Thus there are systems whose states are reliable indicators of earlier conditions. However, there is no analogous reasons to expect the existence of 'prerecording systems', for there is no time reverse of V-correlation.

Probably the most significant example of the knowledge asymmetry is that we can remember the past but have no analogous faculty with respect to the future. Anticipation is not as trustworthy as recollection, and this difference is reflected in a phenomenological difference between them, which in turn plays an essential role in our experience of the passage of time. To put the matter crudely, our concept of time is dependent on our capacity to observe that one event is earlier than another. This ability depends on our remembering one of the events while perceiving the other. And this phenomenon is an instance of the knowledge asymmetry. Thus, in explaining the knowledge asymmetry, we are uncovering a source of our conception of time order.

The concept of time order enters into our notion of explanation. For explanation, in any domain, is a matter of economical systematization—that is to say, a matter of deducing all the facts in the domain from a basic core of principles that, when formulated in 'natural' terms, are as simple as possible. In the special context of physical phenomena it turns out that the way to do this is with a combination of 'naturally simple' generalizations and a body of particular facts. As the most basic body of particular facts we pick the initial conditions of the universe: we prefer to explain in terms of *earlier* rather than *later* phenomena. This time bias is sustained by our attachment to a cluster of crude but entrenched maxims that partially constitute our concept of explanation: for example, that what we know explains what we don't know, that correlations between events are explainable, and that rational decisions are made for the sake of events whose occurrence they might explain. Each of these maxims, when conjoined with further beliefs, independently entails that explanation be given in terms of preceding circumstances. Thus the first maxim may be combined with our knowledge of history and our ignorance of the future, the second goes with the fork asymmetry, and the third may be linked with the view that rational choice is usually oriented toward the future. Taken together, these considerations show why time asymmetry is a well-entrenched feature of event explanation.

Having seen why explanation tends to be future oriented, it is then possible to understand the time bias in our conception of causation. For causation is the relation we attribute to pairs of events when the fact describing one of the events is cited to explain the fact describing the other. So from the direction of explanation we may infer the direction of causation. Notice, however, that what is accounted for here is merely why one should *hold* that causes typically precede their effects. In order to explain the *fact* of causal time asymmetry, it is necessary to appeal to the structure of the world itself, and not just to our beliefs about it. To that end, I proposed a neo-Humean theory according to which the causal relation is composed, in a complex way, from chains of direct, nomological determination and time order. Thus it turns out to be *constitutive* of causation that causes typically precede their effects.

This theory denies the reality—but not the conceivability—of local backward causation. In other words, one can accept that there are *hypothetical* evidential circumstances in which such a thing should be postulated. However, in *actual* cases (e.g., Feynmann's conception of the position as an electron moving backward in time) the theoretical advantages of postulating backward causation are outweighed by disadvantages. Not the least of these undesirable consequences would be the need to give up a very simple theory of causation (with no alternative in sight) and to abandon the principle that correlated events are causally connected. For such reasons we should not, at present, accept the existence of local backward causation.

However, this position does not detract from the plausibility of *global* backward causation of the sort that would be involved in Gödelian time travel. That prospect depends on whether spacetime has the right structure. Even if it does, the arguments about 'murdering one's great-grandfather' show that we would not be able to visit our own history. But we could still travel to the spatially distant past, from where it would not be possible to engage in 'self-defeating' activities. Thus, a form of backwards causation could take place.

It is often said that counterfactual dependence is time-asymmetric: if the past had been different, then the future would be, but not the other way round. This alleged asymmetry is a central component of Lewis's approach to the issues we have been discussing. For he defines causation in terms of counterfactual dependence and derives the direction of causation from the direction of counterfactual dependence. However, we have found there to be many objections to his definition, to the theory of counterfactuals on which it is based, and to the thesis that counterfactual dependence is time-asymmetric. These problems can be avoided, while preserving the idea that causation and counterfactuals are intimately related, by inverting Lewis's direction of analysis. With an account of causation already in hand, we are free to employ causal notions in our analysis of counterfactuals. We can take an approach along the lines proposed by Goodman, according to which a counterfactual holds when the truth of its antecedent entails, given prevailing conditions, the truth of its consequent. And we can characterize 'the prevailing conditions' as those circumstances that are independent—logically, nomologically, and *causally*—of the falsity of the antecedent. Thus we are able to explain why, for example, it is not true that if a certain unstruck match *had* been struck, there would have been no oxygen. For, although that consequent is entailed by the antecedent, given the actual nonlighting of the match, and so on, we are not allowed to include the nonlighting among prevailing conditions, since it is an effect of the falsity of the antecedent.

Returning to the knowledge asymmetry, we have seen how it may help to explain our conception of time order, and thereby the direction of explanation and causation. In addition the knowledge asymmetry is at the basis of our experience of free choice. For the feeling of freedom is associated with our ignorance, prior to deliberation, of what we are going to do. If we knew as much about the future as we do about the past (and with the same certainty that is provided by memory) this feeling could not be sustained. We can then explain, in terms of the sense of free choice, why rational decisions are not made for the sake of past events and why we care about the future more than the past.

More specifically, the sense of free choice is the conviction that one's actions will be the product of a process of deliberation in which the recognition of one's beliefs and desires form the basis of reasoning that lead to a conclusion about what will be done. Thus Wittgenstein (1922, sec. 1.362) was right to stress that freedom implies not knowing in advance what one is going to do; such knowledge must emerge through deliberation. But he was wrong, I think, to suggest that freedom *consists* in this epistemological fact. Unpredictable, uncontrollable movements might well not be free. Rather, the essence of freedom is the role of deliberation, and the absence of prior knowledge of what will be done is a consequence of this.

Given that free action involves deliberation, it seems clear why, in the normal course of things, actions will not be motivated by a desire for the occurrence of past events. For when an agent contemplates a particular act, there is nothing he can infer about the probability (relative to that act) of the desired past event that he cannot already infer from the knowledge he already has of his own beliefs and desires.

This is because his beliefs and desires mediate any causal connection between the past and the act in question. So the only correlations between past conditions and future actions of which one is normally aware are derived from correlations between those past conditions and one's present motivational state. Therefore the present motivational state is a kind of 'probabilistic screen'. For any past event E and any pair of alternative actions A and B, the agent's subjective probability will satisfy

$$P(E/A) = P(E/B)$$

Therefore E can provide no motive to do A rather than B, or vice versa. This situation will change only in bizarre circumstances, like Newcomb's problem, where it is stipulated that the agent's beliefs and desires do not act as screens.

The respect in which we care more about the future than the past is that we want the satisfaction of our future selfish desires much more than we want the satisfaction of our past ones. Whether or not this is rational, I do not discuss. But I believe it is and, moreover, that its rationality does not require the existence of any explicit rationale. There remains the question of explanation. And this, it seems plausible to suppose, is a matter of selectional value. It pays to be concerned about one's future selfish desires because that concern will increase the chances of their satisfaction, and their satisfaction would be beneficial. On the other hand, there is no point in strongly wishing for the satisfaction of past desires. It wouldn't do any good because such wishes are not correlated with their fulfillment.

Finally, there is the direction of time itself. At first sight the existence of so many pervasive, temporally asymmetric phenomena might tempt one to think that time itself must have some intrinsic directional character. The hope would be that this fundamental asymmetry would account for all the others in a unified way. However, if what I have been saying is along the right lines, we might be able to give a different total explanation of asymmetries in time—an explanation based ultimately on the fork asymmetry and its cosmological sources. In that case there would be little reason to suppose that time itself is asymmetric.

The air of certainty and lack of qualification that mark the last few pages may be justified only by the desirability of providing a compact account of what has been done here. True, nothing has been asserted unless there are supporting arguments somewhere in the text. However, I am afraid there are many theses that never receive proper clarification and defense. I should call attention to the following especially egregious loose ends:

1. *Verificationism.* The 'moving *now*' conception of time relies, as we saw in chapter 2, on the thesis that there is a variation in facts from one moment of time to another. And this thesis depends in turn on verificationism. For obviously one's *view* of the facts varies. But it is only from the perspective of a verificationist account of meaning that one would be tempted to suppose that such changes in evidential circumstances entail changes in the facts themselves. Thus it seems that a refutation of the 'moving *now*' conception of time hinges on a refutation of verificationism. But, beyond some hints about the way this argument might proceed, such a project could not be conducted here.

2. *Natural properties.* This murky, notion appears throughout philosophy and badly stands in need of explanation. It was relied on here, at a couple of points. For example, I defined the anisotropy (i.e., asymmetry) of time as the presence of an *intrinsic* difference between the past and future directions. The intrinsic properties of a thing are intuitively those that make no reference to other things. And such reference may normally be detected in the syntactic structure of predicates. However, this will be so only if the language is *natural*—only if primitive predicates refer to natural kinds such as red, round, and ruby and not grue or grandmother. But what makes a property 'natural'?

A second use for the notion appeared in chapter 10 where we wished to characterize the idea of projectibility and the concept of law. Among all the hypotheses that fit our data we tend to rely on (project) those that, roughly speaking, 'look simple'. Now, simplicity can normally be judged on syntactic grounds. But, once again, this will be so only if bizarre redefinitions are excluded, that is, if the language is 'natural'. Thus the explications of 'projectibility' and 'law' seem to depend on the idea of 'natural property':

3. *Projectible hypotheses.* Further problems having to do with projectibility concern not so much its definition, as its application. Thus, in chapter 3, I argued that time is probably isotropic, on the grounds that no anisotropy has been discovered so far. This is, in effect, to assume that 'All laws are time-symmetric" is a projectible hypothesis—which seems true to me, although, admittedly, not obviously true. Perhaps a better understanding of projectibility would help to settle the matter.

Similarly there is a specific question of projectibility in the discussion of time travel. I maintained that the absence, in our experience, of certain kinds of coincidence counts against the possibility of trips into the local past. For that possibility, I claimed, would imply an uncaused correlation between taking such trips and not trying to 'change the past' in certain ways. This reasoning assumes that the scarcity of coincidences, which we now observe, would remain in effect in the circumstances of time travel. And although this projection seems legitimate to me, the issue deserves further attention.

4. *Entropy.* My aim in chapter 4 was to give a nontechnical account of some of the problems facing mechanical explanations of *de facto* one-way processes, given that the basic laws of nature are time-symmetric. In view of these problems I tried to say something very informally about the shape that an adequate explanation would have to take. I have not tried to present a physical theory but, rather, a speculative, idealized picture. Nevertheless, any account, however crude, must in the end be responsible to the known facts and to our best scientific theories. Consequently it is important to clarify the concepts employed—specifically, the notions of 'cosmic input noise' and 'random initial conditions'—to the point at which their causal properties may be detailed and their consonance with scientific knowledge may be fully evaluated.

In the light of such clarification it should be possible to give a more exact description and explanation of the principle of V-correlation. No one would deny, I imagine, that this principle embodies an important truth. It is related to the fact that correlated events are causally connected. However, a precise statement of the principle is hard to give. And this is because the sources (and hence the limitations) of the principle are not fully understood. I have tried to sketch its dependence on the 'microscopic randomness of initial conditions'. I would hope that a clear statement of that condition would yield a better explanation.

The handful of problems I have just mentioned are merely the most glaring of the respects in which this book is incomplete. Clearly I have not been able to do justice to all the topics that have arisen, and the reader will have found many places at which further discussion would have been desirable. It seemed to me, however, that the basic structure of a theory of causation, explanation—or any of the other things at issue here—should be investigated by making sure that it meshes with accounts of neighbouring concepts; and that, unless this coherence is made plausible, there is little point in struggling to work out the details. Therefore I hope that some of the deficiencies of this essay can be seen as reasonable sacrifices for the sake of getting a harmonious overall picture of temporally asymmetric phenomena.

Bibliography

Alexander, H. G. (ed.). 1956. *The Leibniz-Clarke Correspondence Manchester.* Manchester: Manchester University Press.

Anscombe, G. E. M. 1971. *Causality and Determination.* Cambridge: Cambridge University Press. Reprinted in *Causation and Conditionals.* Edited by E. Sosa. Oxford: Oxford University Press. 1975.

Arnold, V. I., and Avez, A. 1968. *Ergodic Problems of Classical Mechanics.* New York: Benjamin.

St. Augustine. *Confessions.* Translated by E. B. Pusey. Chicago: Henry Regnery Co., 1948.

Bennett, J. 1984. "Counterfactuals and Temporal Direction." *Philsophical Review* 93:57–91.

Birkhoff, G. D. 1931. "Proof of the Ergodic Theorem." *Proceedings of the National Academy of Sciences* 17:656–660.

Black, M. 1956. "Why Cannot an Effect Precede Its Cause." *Analysis* 16:49–58.

Black, M. 1959. "The Direction of Time." Reprinted in his Models and Metaphors, Ithaca: Cornell University Press. 1962.

Blatt, J. M. 1959. "An Alternative Approach to the Ergodic Problem." *Progress in Theoretical Physics* 22:745–756.

Boltzmann, L. 1898. *Lectures on Gas Theory.* Translated by S. G. Brush. Berkeley: University of California Press, 1964.

Braithwaite, R. B. 1927/8). "The Idea of Necessary Connexion." *Mind* 36:467–77; *Mind* 37:62–72.

Braithwaite, R. B. 1953. *Scientific Explanation.* Cambridge: Cambridge University Press.

Broad, C. D. 1938. *Examination of McTaggart's Philosophy.* Cambridge: Cambridge University Press.

Bromberger, S. 1966. "Why-Questions." *Mind and Cosmos.* Vol. 3. University of Pittsburgh Series in the Philosophy of Science.

Brush, S. 1966. *Kinetic Theory.* Vol. 2 Oxford: Pergamon.

Cartwright, N. 1979. "Causal Laws and Effective Strategies." *Nous* 419–438. Reprinted in *How the Laws of Physics Lie.* Oxford: Clarendon Press, 1983.

Chisholm, R. M. 1955. "Law Statements and Counterfactual Inference." *Analysis* 15:97–105. Reprinted in *Causation and Conditionals.* Edited by E. Sosa. Oxford: Oxford University Press, 1975.

Chisholm, R. M., and Taylor, R. 1960. "Making Things to Have Happened." *Analysis* 20:73–78.

Cocke, W. J. 1967. "Statistical Time Symmetry and Two-Time Boundary Conditions in Physics and Cosmology." *Physical Review* 160:1165–1170.

Collingwood, R. G. 1940. *An Essay on Metaphysics.* Oxford: Clarendon Press.

Davidson, D. 1967. "Causual Relations." *Journal of Philosophy* 64:691–703. Reprinted in *Causation and Conditionals*. Edited by E. Sosa. Oxford: Oxford University Press, 1975.

Davies, P. C. W. 1974. *The Physics of Time Asymmetry*. Berkeley: University of California Press.

Dretske, F. I. 1972. "Contrastive Statements." *Philosophical Review* 81:411–437.

Dummett, M. 1954. "Can an Effect Precede Its Cause." *Proceedings of the Aristotelian Society Supplementary* 28:27–44.

Dummett, M. 1960. "A Defense of McTaggart's Proof of the Unreality of Time." *Philosophical Review* 69:497–504.

Dummett, M. 1964. "Bringing about the Past." *Philosopical Review* 73:338–359.

Dummett, M. 1976. *Truth and Other Enigmas*. Oxford: Clarendon Press.

Earman, J. 1967. "Irreversibility and Temporal Asymmetry." *Journal of Philosophy*, 64:543–549.

Earman, J. 1969. "The Anisotropy of Time." *Australasian Journal of Philosophy* 47:273–295.

Earman, J. 1972. "Implications of Propagation Outside the Null-Cone." *Australasian Journal of Philosophy* 50:223–237.

Earman, J. 1974. "An attempt to Add and Little Direction to 'The Problem of the Direction of Time.'" *Philosophy of Science*. 41:15–47.

Earman, J. 1976. "Causation: A Matter of Life and Death." *Journal of Philosophy*. 73:5–25.

Eells, E. 1982. *Rational Decision and Causality*. Cambridge: Cambridge University Press.

Ehrenfest, P., and Ehrenfest, T. 1959. *The Conceptual Foundation of the Statistical Approach to Mechanics*. Translated by M. Moravcsik. Ithaca: Cornell University Press.

Feynman, P. R. 1949. "The Theory of Positrons." *Physics Review* 76:749–759.

Flew, A. 1954. "Can an Effect Precede Its Cause." *Proceedings of the Aristotelian Society* (Supplementary volume) 38:45–62.

Flew, A. 1956. "Effects Before Their Causes? Addenda and Corrigenda", Analysis 16:104–110.

Flew, A. 1957. "Causal Disorder Again", Analysis 17:81–86.

Friedman, M. 1974. "Explanation and Scientific Understanding." *Journal of Philosophy* 71:5–19.

Gale, R. 1967. *The Philosophy of Time*. New York: Anchor Books

Gale, R. 1969. *The Language of Time*. London: Routledge and Kegan Paul.

Geach, P. T. 1972. *Logic Matters*. Berkeley: University of California Press

Gettier, E. L. 1963. "Is Justified True Belief Knowledge?" Analysis 23:121–123.

Gasking, D. 1955. "Causation and Recipes." *Mind* 64:474–487.

Gibbard, A., and Harper, W. 1978. "Counterfactuals and Two Kinds of Expected Utility." In *Foundations and Applications of Decision Theory*. Edited by C. Hooker et al. Western Ontario Series in the Philosophy of Science. Vol. 13. Dordrecht: Reidel. Reprinted in *Ifs*. Edited by W. L. Harper et al. Western Ontario Series in Philosophy of Science. Vol. 15. Dordrecht: Reidel, 1980.

Gödel, K. 1949. "A Remark about the Relationship between Relativity Theory and Idealistic Philosophy." *Albert Einstein: Philosopher-Scientist*. Edited by P. Schilpp. La Salle: Open Court, pp. 557–562.

Gold, T. 1962. "The Arrow of Time." *American Journal of Physics* 30:403–410.

Goldman, A. I. 1967. "A Causal Theory of Knowing." Journal of Philosophy 64:355–372.

Goodman, N. 1947. "The Problem of Counterfactual Conditions." *Journal of Philosophy* 44:113–128. Reprinted in *Fact. Fiction and Forecast*. 3rd ed. Indianapolis: Bobbs-Merrill, 1955.

Gould, S. T., & Lewontin, R. C. 1979. "The Spandrels of San Marco and the Panglossian Paradigm—A Critique of the Adaptionist Programme." Proceedings of the Royal Society, London, B205:581–598.

Grünbaum, A. 1963. *Philosophical Problems of Space and Time*. New York: Knopf: 2nd ed., Dordrecht: Reidel, 1973.

Healey, R. 1981. "Statistical Theories, Quantum Mechanics and the Directedness of Time." In *Time and Reduction*. Edited by R. Healey. Cambridge: Cambridge University Press.

Healey, R. 1983. "Temporal and Causal Asymmetry." In *Space, Time and Causality*. Edited by R. Swinburne. Royal Institute of Philosophy Conferences Volume. Dordrecht: Reidel, pp. 79–105.

Hempel, C. G., and Oppenheim, P. 1948. "Studies in the Logic of Explanation." *Philosophy of Science* 15:135–175.

Hempel, C. G. 1965. *Aspects of Scientific Explanation*. New York: The Free Press.

Hempel, C. G. 1966. *Philosophy of Natural Science*. Englewood Cliffs, N. J.: Prentice-Hall.

Horwich P. G. 1975. "On Some Alleged Paradoxes of Time Travel." *Journal of Philosophy* 72:432–444.

Horwich, P. G. 1982. *Probability and Evidence*. Cambridge: Cambridge University Press.

Horwich, P. G. 1985. "Decision Theory in the Light of Newcomb's Problem." *Philosophy of Science*. 52:431–450.

Husserl, E. 1928. *The Phenomenology of Internal Time-Consciousness*. Translated by J. S. Churchill. Bloomington: University of Indiana Press.

Jackson, F. 1977. "A Causal Theory of Counterfactuals." *Australasian Journal of Philosophy* 55:3–21.

Jeffrey, R. 1965. *The Logic of Decision*. New York: McGraw-Hill.

Jeffrey, R. 1981. "The Logic of Decision Defended." *Synthese* 48:473–492.

Jeffrey, R. 1983. *The Logic of Decision*. 2nd ed., Chicago: University of Chicago Press.

Katz, B. D. 1983. "The Identity of Indiscenibles Revisited." *Philosophical Studies* 44:37–44.

Kim, J. 1973. "Causes and Counterfactuals." *Journal of Philosophy* 70:570–572. Reprinted in *Causes and Conditionals*. Edited by E. Sosa. Oxford: Oxford University Press, 1975.

Klein, M. 1973. "The Development of Boltzmann's Statistical Ideas." *The Boltzmann Equation—Theory and Applications*. Edited by E. G. D. Cohen and W. Thirring. Acta Physica Austriaica Suppl. 10. New York: Springer, pp. 53–106.

Kuhn, T. S. 1978. *Black-Body Theory and the Quantum Discontinuity 1894–1912*. Oxford: Clarendon Press.

Layzer, D. 1975. "The Arrow of Time." *Scientific American* 234:56–69.

Lebowitz, J. A., and Bergmann, P. G. 1955. "A New Approach to Non-Equilibrium Processes." *Physical Review* 99:578–587.

Levi, I. 1975. "Newcomb's Many Problems." *Theory and Decision* 6:161–175.

Lewis, D. 1973a. *Counterfactuals*. Oxford: Blackwell.

Lewis, D. 1973b. "Causation." *Journal of Philosophy* 10:556–567.

Lewis, D. 1976. "The Paradoxes of Time Travel." *American Philosophical Quarterly* 13. 145–152.

Lewis, D. 1979a. "Prisoner's Dilemma Is a Newcomb Problem." *Philosophy and Public Affairs* 8:235–240.

Lewis, D. 1979b. "Counterfactual Dependence and Time's Arrow." *Nous* 13:455–476.

Lewis, D. 1981. "Causal Decision Theory." *Australasian Journal of Philosophy* 59:5–30.

Lewis, D. 1983. "New Work for a Theory of Universals." *Australasion Journal of Philosophy* 61:343–377.

Lewis, D. 1987. *Collected Papers*. Vol. 2. Oxford: Oxford University Press.

Loschmidt, J. 1876. "Über den Zustand des Warmegleichgewichtes eines Systems von Köpern mit Rücksicht auf die Schwerkraft." *Sitzungsberichte der Kaiserlich der Wissenschaften in Wien, Mathematisch-Naturwissenschaflichen* 73:139.

Mackie, J. L. 1974. *Causation: The Cement of the Universe*. Oxford: Oxford University Press.

Malament, D. 1985a. "'Time Travel' in the Godel Universe." *Proceedings of the Philosophy of Science Association 1984 Meetings*, Vol. 2.

Malament, D. 1985b. "Minimal Acceleration Requirements for Time Travel in Godel Space-Time." *Journal of Mathematical Physics* 26:774–777.

Matthews, G. 1979. "Time's Arrow and the Structure of Spacetime." *Philosophy of Science* 46:82–97.

Mehlberg, H. 1961. "Physical Laws and Time's Arrow." *Current Issues in the Philosophy of Science*. Edited by H. Feigl and G. Maxwell. New York: Holt, Rinehart, and Winston.

Mehlberg, H. 1962. Review of *The Direction of Time* by H. Reichenbach, Philosophical Review 71:99–104.

Mellor, D. H. 1981. *Real Time*. Cambridge: Cambridge University Press.

Miller, I. 1984. *Husserl, Perception, and the Awareness of Time*. Cambridge, Mass.: The MIT Press. A Bradford Book.

Morrison, P. 1966. "Time's Arrow and External Perturbations." In *Preludes in Theoretical Physics*. Edited by A. De Shalit. Amsterdam: North Holland.

Newton-Smith, W. H. 1980. *The Structure of Time*. London: Routledge and Kegan Paul.

Nozick, R. 1969. "Newcomb's Problem and Two Principles of Choice." In *Essays in Honor of Carl G. Hempel*. Edited by N. Rescher. Dordrecht: Reidel.

Papineau, D. 1985. "Causal Asymmetry." *British Journal for the Philosophy of Science* 36:273–289.

Parfit, D. 1985. *Reasons and Persons*. Oxford: Oxford University Press.

Pears, D. 1957. "The Priority of Causes," *Analysis* 17:54–63.

Popper. K. R. 1956/7/8. "The Arrow of Time" *Nature* 177:538; 179:382; 179:1297; and 181:404.

Popper, K. R. 1959. *The Logic of Scientific Discovery*. London: Hutchinson.

Popper, K. R. 1972. *Objective Knowledge*. Oxford: Oxford University Press.

Prior, A. 1967. *Past, Present and Future*. Oxford: Oxford University Press.

Putnam, H. 1962. "The Analytic and the Synthetic." In Minnesota Studies in the Philosophy of Science Vol. III. Edited by H. Feigl and G. Maxwell. Minnesota: University of Minnesota Press.

Putnam, H. 1978. *Meaning and the Moral Sciences*. London: Routledge and Kegan Paul.

Quine, W. V. O. 1951. "Two Dogmas of Empiricism." Reprinted in *From a Logical Point of View*. New York: Harper and Row, 1963.

Quine, W. V. O. 1969. "Natural Kinds." *Ontological Relativity and Other Essays*. New York: Columbia University Press.

Redhead, M. 1983. "Nonlocality and Peaceful Coexistence." In *Space Time and Causality*. Edited by R. Swinburne. Dordrecht: Reidel.

Reichenbach, H. 1956. *The Direction of Time*. Berkeley: University of California Press.

Richter, R. 1984. "Rationality Revisited." *Australasian Journal of Philosophy* 62: 392–403.

Rosenfeld, L. 1955. "On the Foundations of Statistical Thermodynamics." *Acta Physica Polonica* 14:9.

Russell, B. 1903. *The Principles of Mathematics*. New York: Norton.

Salmon, W. C. 1971. *Statistical Explanation and Statistical Relevance*. Pittsburgh: University of Pittsburgh Press.

Salmon, W. C. 1978. "Why Ask 'Why'?" In *Proceedings and Addresses of the American-Philosophical Association* 51:683–705.

Salmon, W. C. 1984. *Scientific Explanation and the Causal Structure of the World*. Princeton: Princeton University Press.

Sanford, D. H. 1976. "The Direction of Causation and the Direction of Conditionship." *Journal of Philosophy* 73:193–207.

Savage, L. J. 1972. *The Foundations of Statistics*. 2nd rev. ed. New York: Dover; 1st, New York: Wiley, 1954.

Schliesinger, G. 1980. *Aspects of Time*. Indianapolis: Hackett Publishing Company.

Schmidt, H. 1966. "Model of an Oscillating Cosmos Which Rejuvenates during Contraction." *Journal of Mathematical Physics*, 494–509.

Schrödinger, E. 1950. "Irreversibility." *Proceedings of the Royal Irish Academy*, vol. 53.

Scriven, M. 1957. "Randomness and the Causal Order." *Analysis* 17:5–9.

Scriven, M. 1964. "An Essential Unpredictability in Human Behavior." In *Scientific Psychology: Principles and Approaches*. Edited by B. B. Wolman and E. Nagel. New York: Basic Books, pp. 411–425.

Scriven, M. 1975. "Causation as Explanation." *Nous* 9:3–16.

Sellars, W. 1962. "Time and the World Order." Minnesota Studies in Philosophy of Science III, University of Minnesota.

Shoemaker, S. S. 1969. "Time without Change." *Journal of Philosophy* 66:363–381.

Shoemaker, S. S. 1980. "Properties, Causation and Projectibility." In *Applications of Inductive Logic*. Edited by L. J. Cohen and M. Hesse. Oxford: Clarendon Press.

Shoemaker, S. S. 1981. "Causality and Properties." In *Time and Cause: Essays Presented to Richard Taylor*. Edited by P. van Inwagen. Dordrecht: Reidel.

Sklar, L. 1974. *Space, Time and Spacetime*. Berkeley: University of California Press.

Sklar, L. 1981. "Up and Down, Left and Right, Past and Future." *Nous* 15:111–129.

Skyrms, B. 1980. *Causal Necessity*. New Haven: Yale University Press.

Slote, M. A. 1978. "Time in Counterfactuals." *Philosophical Review* 87:3–27.

Smart, J. J. C. 1955. "Spatializing Time." Reprinted in *The Philosophy of Time*. Edited by R. Gale. New York: Anchor Books, 1967.

Smart, J. J. C. 1963. *Philosophy and Scientific Realism*. New York: Humanities Press.

Smart, J. J. C. 1967. "Time." *Encyclopedia of Philosophy*. Vol. 8. New York: Macmillan.

Smart, J. J. C. 1968. *Between Science and Philosophy*. New York: Random House.

Smart, J. J. C. 1980. "Time and Becoming." In *Time and Cause*. Edited by P. Inwagen. Dordrecht: Reidel.

Sobel, J. H. 1979. "Probability, Chance and Choice." Unpublished manuscript, University of Toronto.

Stalnaker, R. 1968. "A Theory of Conditionals." In *Studies in Logical Theory*. Edited by N. Rescher. Blackwell APQ Monograph. Oxford: Blackwell. Reprinted in *Causes and Conditionals*. Edited by E. Sosa. Oxford: Oxford University Press, 1975.

Stalnaker, R. 1980. Letter to David Lewis. In *Ifs*. Edited by W. L. Harper et al. Dordrecht: Reidel. Letter written in 1972.

Stalnaker, R. 1984. *Inquiry*. Cambridge, Mass: The MIT Press. A Bradford Book.

Thomson, J. J. 1977. *Acts and Other Events*. Ithaca: Cornell University Press.

Tolman, R. C. 1917. *The Theory of Relativity of Motion*. Berkeley: University of California Press.

Tolman, R. C. 1934. *Relativity, Thermodynamics and Cosmology*. Oxford: Oxford University Press.

van Fraassen, B. C. 1980. *The Scientific Image*. Oxford: Clarendon Press.

van Inwagen, P. (ed.). 1980. *Time and Cause: Essays Presented to Richard Taylor*. Dordrecht: Reidel.

von Neumann, J. 1932. "Proof of the Quasi-Ergodic Hypothesis." *Proceedings of the National Academy of Sciences* 18:70–82.

von Wright, G. H. 1971. *Explanation and Understanding*. Ithaca: Cornell University Press.

Watanabe, S. 1953. "Réversibilité contre irréversibilité en physique quantique." *Louis de Broglie, physicien et penseur*. Paris: Albin Michel.

Weingard, R. 1977. "Space Time and the Direction of Time." *Nous* 11:119–132.

Williams, D. C. 1951. "The Myth of Passage." *Journal of Philosophy* 48:457–472.

Wittgenstein, L. 1922. Tractatus Logico-Philosophicus. London: Routledge & Kegan Paul.

Wittgenstein, L. 1953. *Philosophical Investigations*. Oxford: Oxford University Press.

Index